I0467869

Analysis of 1997–2008 Groundwater Level Changes in the Upper Deschutes Basin, Central Oregon

By Marshall W. Gannett and Kenneth E. Lite, Jr.

Prepared in cooperation with the Oregon Water Resources Department

Scientific Investigations Report 2013–5092

U.S. Department of the Interior
U.S. Geological Survey

U.S. Department of the Interior
SALLY JEWELL, Secretary

U.S. Geological Survey
Suzette M. Kimball, Acting Director

U.S. Geological Survey, Reston, Virginia: 2013

For more information on the USGS—the Federal source for science about the Earth, its natural and living resources, natural hazards, and the environment, visit http://www.usgs.gov or call 1–888–ASK–USGS.

For an overview of USGS information products, including maps, imagery, and publications, visit http://www.usgs.gov/pubprod

To order this and other USGS information products, visit http://store.usgs.gov

Any use of trade, firm, or product names is for descriptive purposes only and does not imply endorsement by the U.S. Government.

Although this information product, for the most part, is in the public domain, it also may contain copyrighted materials as noted in the text. Permission to reproduce copyrighted items must be secured from the copyright owner.

Suggested citation:
Gannett, M.W., and Lite, K.E., Jr., 2013, Analysis of 1997–2008 groundwater level changes in the upper Deschutes Basin, Central Oregon: U.S. Geological Survey Scientific Investigations Report 2013-5092, 34 p., http://pubs.usgs.gov/sir/2013/5092.

Contents

Figures

Figures—Continued

Figures—Continued

Conversion Factors, Datums, and Location System

Conversion Factors

Inch/Pound to SI

Multiply	By	To obtain
Length		
foot (ft)	0.3048	meter (m)
mile (mi)	1.609	kilometer (km)
Area		
acre	4,047	square meter (m^2)
acre	0.4047	square hectometer (hm^2)
square mile (mi^2)	259.0	hectare (ha)
square mile (mi^2)	2.590	square kilometer (km^2)
Volume		
cubic foot (ft^3)	28.32	cubic decimeter (dm^3)
cubic foot (ft^3)	0.02832	cubic meter (m^3)
acre-foot (acre-ft)	1,233	cubic meter (m^3)
acre-foot (acre-ft)	0.001233	cubic hectometer (hm^3)
Flow rate		
acre-foot per year (acre-ft/yr)	1,233	cubic meter per year (m^3/yr)
acre-foot per year (acre-ft/yr)	0.001233	cubic hectometer per year (hm^3/yr)
cubic foot per second (ft^3/s)	0.02832	cubic meter per second (m^3/s)

Temperature in degrees Celsius (°C) may be converted to degrees Fahrenheit (°F) as follows:

$$°F=(1.8\times°C)+32$$

Temperature in degrees Fahrenheit (°F) may be converted to degrees Celsius (°C) as follows:

$$°C=(°F-32)/1.8$$

Conversion Factors, Datums, and Location System

Datums

Vertical coordinate information is referenced to the National Geodetic Vertical Datum of 1929 (NGVD 29).

Horizontal coordinate information is referenced to the North American Datum of 1927 (NAD 27).

Elevation, as used in this report, refers to distance above the vertical datum (NGVD 29).

Location System

The system used for locating wells, springs, and surface-water sites in this report is based on the rectangular system for subdivision of public land. The State of Oregon is divided into townships of 36 square miles numbered according to their location relative to the east-west Willamette baseline and a north-south Willamette meridian. The position of a township is given by its north-south "Township" position relative to the baseline and its east-west "Range" position relative to the meridian. Each township is divided into 36 one-square-mile (640-acre) sections numbered from 1 to 36. For example, a well designated as 14S/16E-32CBA is located in Township 14 south, Range 16 east, section 32. The letters following the section number correspond to the location within the section; the first letter (C) identifies the quarter section (160 acres); the second letter (B) identifies the quarter-quarter section (40 acres); and the third letter (A) identifies the quarter-quarter-quarter section (10 acres). Therefore, well 32CBA is located in the NE quarter of the NW quarter of the SW quarter of section 32. When more than one designated well occurs in the quarter-quarter-quarter section, a serial number is appended.

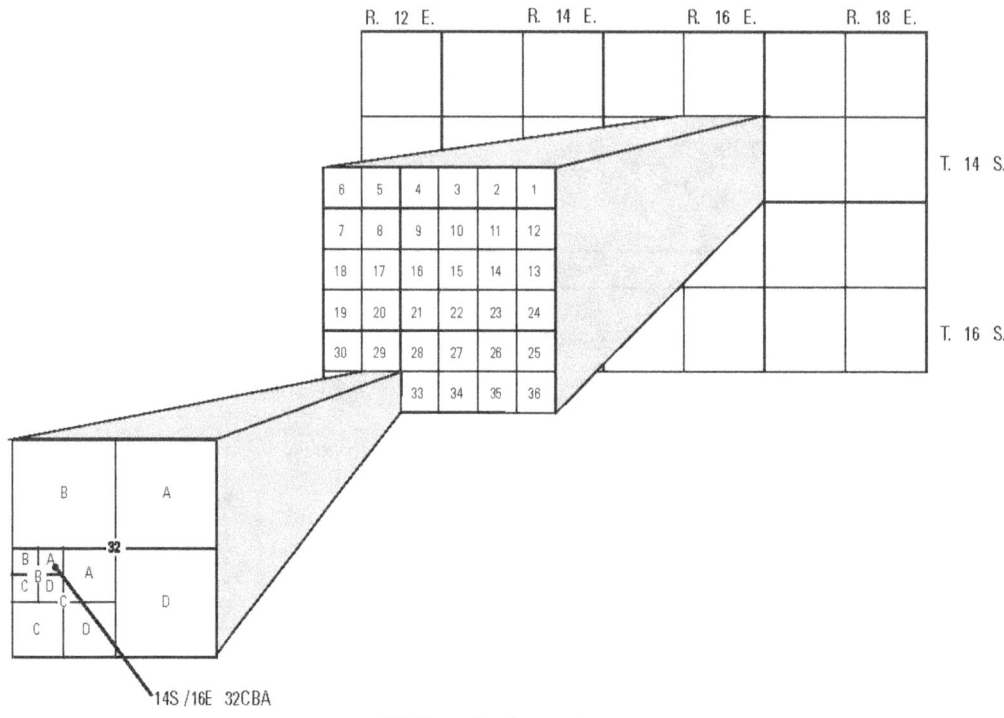

Well-numbering system

Analysis of 1997–2008 Groundwater Level Changes in the Upper Deschutes Basin, Central Oregon

By Marshall W. Gannett and Kenneth E. Lite, Jr.

Abstract

Groundwater-level monitoring in the upper Deschutes Basin of central Oregon from 1997 to 2008 shows water-level declines in some places that are larger than might be expected from climate variations alone, raising questions regarding the influence of groundwater pumping, canal lining (which decreases recharge), and other human influences. Between the mid-1990s and mid-2000s, water levels in the central part of the basin near Redmond steadily declined as much as 14 feet. Water levels in the Cascade Range, in contrast, rose more than 20 feet from the mid-1990s to about 2000, and then declined into the mid-2000s, with little or no net change.

An existing U.S. Geological Survey regional groundwater-flow model was used to gain insights into groundwater-level changes from 1997 to 2008, and to determine the relative influence of climate, groundwater pumping, and irrigation canal lining on observed water-level trends. To utilize the model, input datasets had to be extended to include post-1997 changes in groundwater pumping, changes in recharge from precipitation, irrigation canal leakage, and deep percolation of applied irrigation water (also known as on-farm loss). Mean annual groundwater recharge from precipitation during the 1999–2008 period was 25 percent less than during the 1979–88 period because of drying climate conditions. This decrease in groundwater recharge is consistent with measured decreases in streamflow and discharge to springs. For example, the mean annual discharge of Fall River, which is a spring-fed stream, decreased 12 percent between the 1979–88 and 1999–2008 periods. Between the mid-1990s and late 2000s, groundwater pumping for public-supply and irrigation uses increased from about 32,500 to 52,000 acre-feet per year, partially because of population growth. Between 1997 and 2008, the rate of recharge from leaking irrigation canals decreased by about 58,000 acre-feet per year as a result of lining and piping of canals. Decreases in recharge from on-farm losses over the past decade were relatively small, approaching an estimated 1,000 acre-feet per year by the late 2000s. All these changes in the hydrologic budget contributed to declines in groundwater levels.

Groundwater flow model simulations indicate that climate variations have the largest influence on groundwater levels throughout the upper Deschutes Basin, and that impacts from pumping and canal lining also contribute but are largely restricted to the central part of the basin that extends north from near Benham Falls to Lower Bridge, and east from Sisters to the community of Powell Butte. Outside of this central area, the water-level response from changes in pumping and irrigation canal leakage cannot be discerned from the larger response to climate-driven changes in recharge. Within this central area, where measured water-level declines have generally ranged from about 5 to 14 feet since the mid- 1990s, climate variations are still the dominant factor influencing groundwater levels, accounting for approximately 60–70 percent of the measured declines. Post-1994 increases in groundwater pumping account for about 20–30 percent of the measured declines in the central part of the basin, depending on location, and decreases in recharge due to canal lining account for about 10 percent of the measured declines. Decreases in recharge from on-farm losses were simulated, but the effects were negligible compared to climate influences, groundwater pumping, and the effects of canal lining and piping.

Observation well data and model simulation results indicate that water levels in the Cascade Range rose and declined tens of feet in response to wet and dry climate cycles over the past two decades. Water levels in the central part of the basin, in contrast, steadily declined during the same period, with the rate of decline lessening during wet periods. This difference is because the water-level response from recharge is damped as water moves (diffuses) from the principal recharge area in the Cascade Range to discharge points along the main stems of the Deschutes, Crooked, and Metolius Rivers in the central part of the basin. Water levels in the central part of the basin respond more to multi-decadal climate trends than shorter term changes.

Groundwater-flow simulations show that the effects from increased pumping and decreased irrigation canal leakage extend south into the Bend area. However, the only wells presently monitored in the Bend area are heavily influenced by the Deschutes River, which dampens any response of water levels to external stresses such as groundwater pumping, changes in canal leakage, or climate variations.

Introduction

Study Area and Previous Work

The upper Deschutes Basin study area spans the part of central Oregon extending eastward from the crest of the Cascade Range to the low-permeability volcanic uplands of the Blue Mountains province (fig. 1). The northern boundary corresponds primarily to the geologic contact between late Tertiary volcanic deposits of Cascade Range origin and older, less permeable, early Tertiary deposits of the John Day Formation. The study area's southern boundary corresponds to the boundary of the Deschutes River drainage. Interior parts of the basin are dominated by a broad volcanic plain punctuated by volcanic eruptive centers, the largest of which is Newberry Volcano. The region is dominated by late Tertiary to Quaternary volcanic deposits that are moderately to highly permeable. Most of the Cascade Range is at elevations greater than 5,000 ft, and major peaks exceed 10,000 ft. The Cascade Range intercepts much of the moisture in eastward-moving air masses from the Pacific Ocean. As a consequence, average precipitation exceeds 75 in/yr over most of the Cascade Range, but decreases to less than 12 in/yr in the central part of the basin.

The combination of high rates of precipitation and highly permeable bedrock results in a large amount of groundwater recharge in the Cascade Range. An estimated 50–70 percent of precipitation infiltrates to the groundwater system in the Cascade Range (Manga, 1997; Gannett and others, 2001). This recharge feeds a substantial regional aquifer system that extends from the Cascade Range to the older volcanic uplands east and north of the upper Deschutes Basin.

The upper Deschutes Basin is drained by the Deschutes River and its many tributaries. Streams in the upper Deschutes Basin are considered fully allocated and closed to additional appropriation. As a consequence, the regional aquifer system has been developed for agricultural and public water supplies. In addition, the vast majority of residents outside of cities depend on wells for domestic water supplies. Rapid growth of the upper Deschutes Basin in the past few decades has relied exclusively on the development of groundwater resources. Maintaining stable and reliable long-term groundwater supplies is critical to the region.

The U.S. Geological Survey conducted a regional groundwater characterization and modeling study in the upper Deschutes Basin in the mid-1990s in cooperation with the Oregon Water Resources Department (OWRD), local government agencies, and the Confederated Tribes of the Warm Springs Reservation (Gannett and others, 2001; Gannett and Lite, 2004). During that study, approximately 1,500 wells were field inventoried. Water levels were monitored in 89 of these wells from 1993 to 1997 (Caldwell and Truini, 1997). Additional water-level data for a subset of these wells were available from an earlier period of monitoring from 1978

to 1980. Analysis of data collected through 1997 indicated that water-level fluctuations were driven primarily by climate cycles. At that time, water-level declines related to groundwater pumping were not apparent in the data.

Post-1997 Water-Level Trends

Rapid population growth and the associated development of the groundwater resources in the upper Deschutes Basin has continued since the late 1990s. Water-level monitoring also has continued in a relatively small subset of wells. Water-level data collected since the late 1990s indicate a continued response to climate cycles in most wells, but many of the wells in the more developed central part of the upper Deschutes Basin appear to show declines larger than what might be expected from climate variations alone. For example, wells close to the Cascade Range and near Sisters in the western part of the upper Deschutes Basin (fig. 2, wells A and B) exhibited (1) drought-related water-level declines from the mid-1980s to about 1995, (2) water-level recovery in 1996 and 1997, (3) fairly stable water levels from 1997 to 2000, (4) another climate-driven decline from 2000 to 2006, and (5) another recovery between 2006 and 2008. The lowest water levels associated with the drought that occurred around 2005 were about the same as, or just slightly lower than, the lowest water levels associated with the previous drought in about 1995. In contrast, water levels around Redmond and the area to the east (fig. 2, wells C and D) showed no recovery since the drought in the mid-1990s, and water levels have been on a more or less steady decline ever since. As of 2008, water levels in wells near Redmond were about 10–14 ft lower than the lowest water levels associated with the previous drought.

There are a number of possible causes of the measured water-level declines in parts of the upper Deschutes Basin. Chief among possible causes are climate-related decreases in basin-wide groundwater recharge from precipitation, increased groundwater pumping, and decreases in local groundwater recharge as a result of lining and piping of irrigation canals. Flow data from groundwater-dominated streams indicate that the Cascade Range aquifers in the upper Deschutes Basin have been affected by a drying trend since the 1950s. This can be seen in the general decline in flow of spring-fed streams in undeveloped parts of the basin. For example, since the 1950s, the annual mean streamflow of Fall River has shown a steady decline superimposed over decadal wet and dry cycles (fig. 3). A general decrease in flow of groundwater-dominated stream systems emanating from the Cascade Range in the Klamath, Rogue, and Umpqua Basins over the past 50 years has been documented by Mayer and Naman (2011). Luce and Holden (2009) documented decreases in streamflow from the Santiam and Metolius Rivers as part of a larger analysis of streams in the Pacific Northwest. Some proportion of the groundwater-level declines in the central parts of the upper Deschutes Basin is the result of this long-term drying trend.

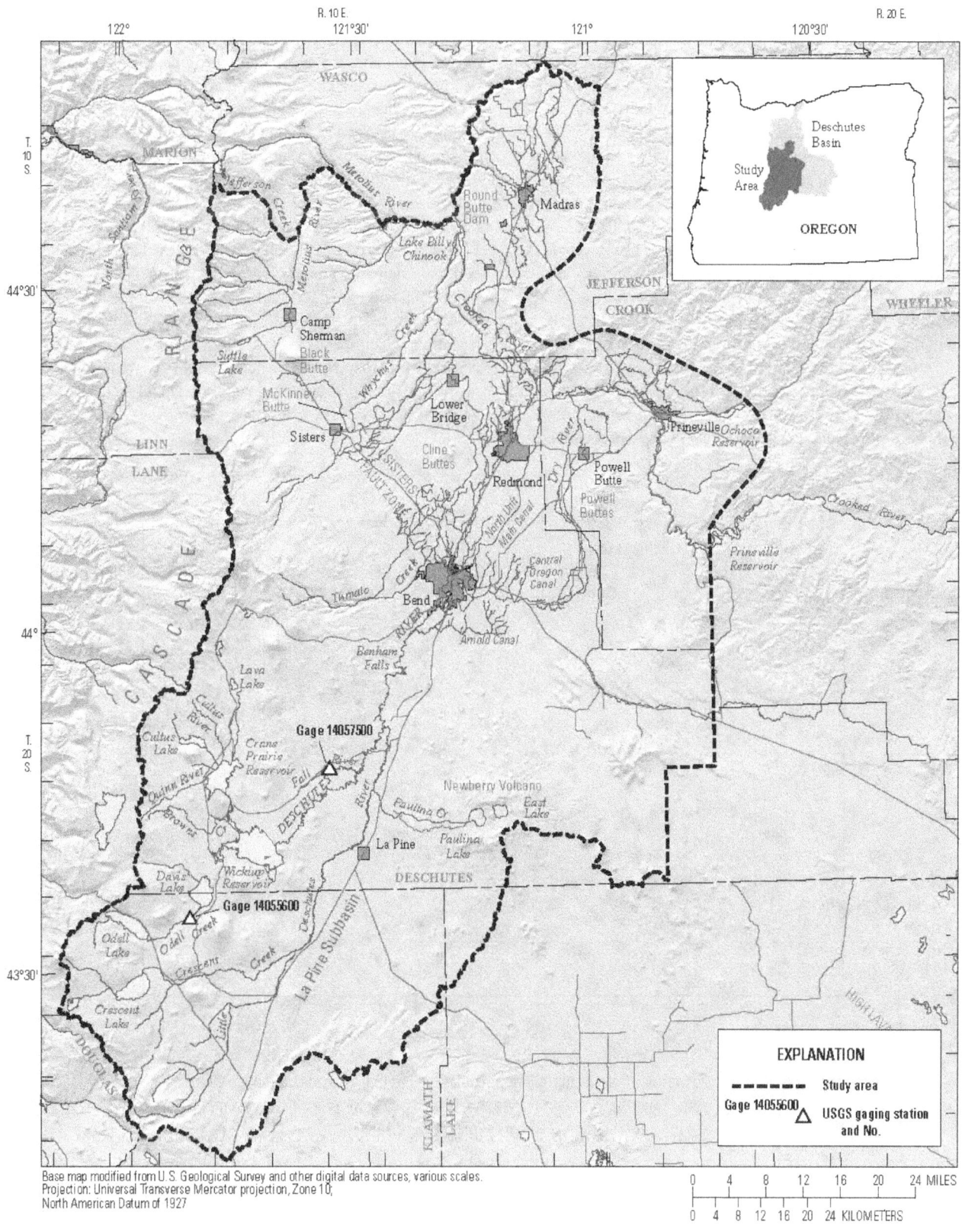

Figure 1.   Major geographic and cultural features of upper Deschutes Basin, central Oregon.

There also are potential human influences on water-level changes in the basin including changes in pumping and irrigation-related activities. Groundwater pumping affects head-distribution in aquifers, and water-level declines are a normal consequence. Generally, if pumping rates do not exceed the aquifer's ability to provide water, water levels will stabilize over time. Groundwater pumping has continued to increase in the basin since the mid-1990s, and the measured declines are at least partially coincident with known pumping centers.

Groundwater recharge from leaking irrigation canals elevated water levels in the central part of the upper Deschutes Basin over the past century. Previous studies (Sceva, 1968; Gannett and others, 2001) have shown that canal leakage is a significant component of the groundwater budget and also has resulted in increased baseflow to the lower Crooked River. Groundwater-level measurements from the early 1900s before installation of the canal network are virtually nonexistent, so it is not possible to know how much water levels have

risen in response to irrigation canal leakage. During the past two decades, there has been substantial lining and piping of irrigation canals for conservation purposes. The consequent decrease in recharge has resulted in declines in groundwater levels in some areas, particularly near the affected canals.

There also is a small amount of artificial groundwater recharge from deep percolation of applied irrigation water. These "on-farm" losses in the upper Deschutes Basin were estimated by Gannett and others (2001) to average about 49,000 acre-ft/yr (68 ft³/s) in the mid-1990s. From 2001 to 2009, on-farm losses decreased slightly because small amounts of acreage have been taken out of production and the water rights have been leased, so water may remain in streams. Part of the reason for doing this is to mitigate the impacts of new groundwater uses. Estimated decreases in groundwater recharge from on-farm losses were included in simulation work for this study, but their effects on groundwater levels are negligible.

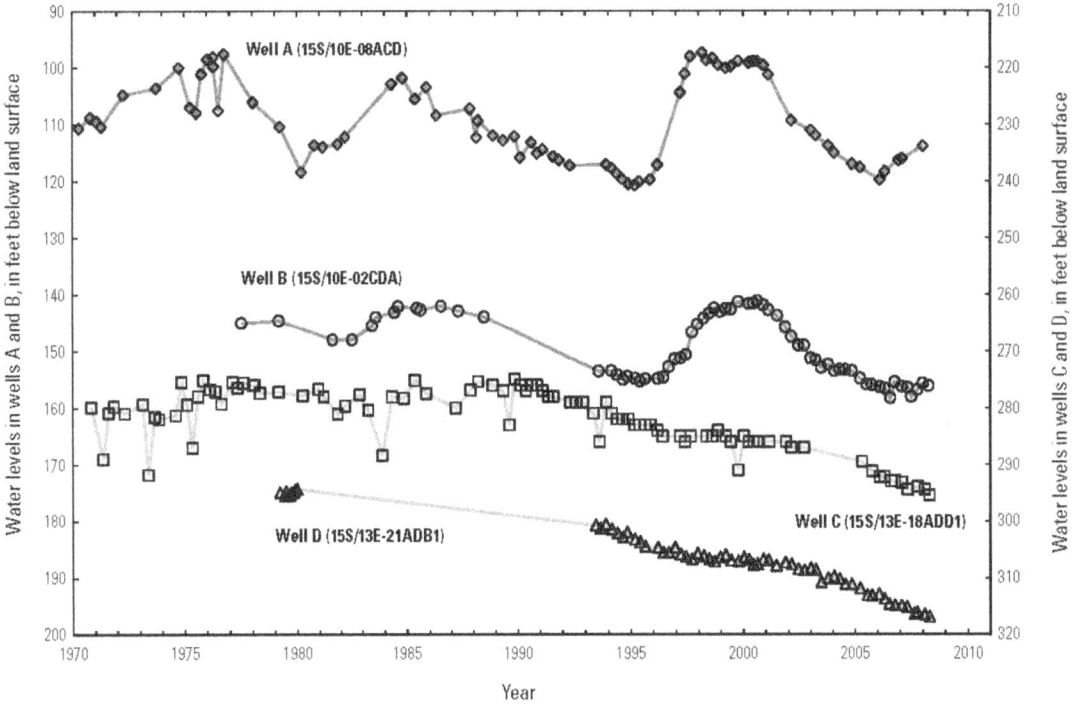

Figure 2.   Selected observation wells in the upper Deschutes Basin, central Oregon, contrasting water-level trends in and near the Cascade Range (wells A and B) with trends in the basin interior (wells C and D). See figure 4 for location of observation wells.

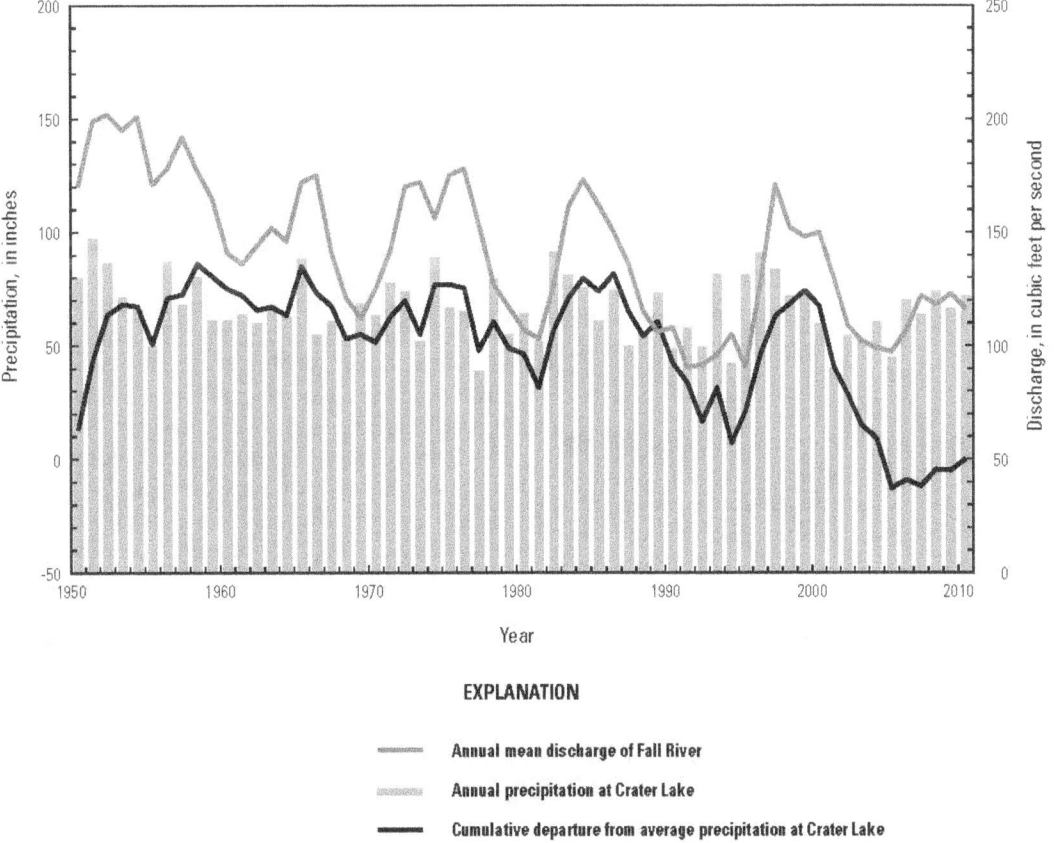

Figure 3. Annual mean discharge of Fall River in the upper Deschutes Basin, central Oregon (USGS gage 14057500); annual (water-year) precipitation at Crater Lake, Oregon; and cumulative departure from average precipitation at Crater Lake. Location of streamflow gaging station is shown in figure 1.

Purpose and Scope

This study was designed to investigate the spatial distribution and causes of measured water-level declines in the upper Deschutes Basin. An important goal was to determine the relative influence of the major probable causes: (1) climate-driven decreases in groundwater recharge, (2) increased groundwater pumping, and (3) decreased artificial recharge as a result of lining and piping of irrigation canals. The study relied largely on historical data and data collected by OWRD between 1997 and 2008 (Oregon Water Resources Department, 2013). As part of the study, some wells measured until 1997 were revisited and measured. No new monitoring efforts have been started specifically as part of this study. The investigation used the groundwater model developed for the upper Deschutes Basin in the 1990s by Gannett and Lite (2004). Input data files were created to allow simulation of conditions up to 2008, but the model was otherwise unchanged.

Methods

This study used the existing USGS upper Deschutes Basin groundwater model (Gannett and Lite, 2004) to evaluate water-level changes between 1997 and 2008. This evaluation entailed compiling data on water-level changes as well as extending model input files to cover the time period of interest.

The upper Deschutes Basin groundwater model encompasses approximately 4,000 mi^2 with a grid composed of 127 rows, 87 columns, and 8 layers with variable cell dimensions. The model was developed using the USGS modular groundwater modeling code MODFLOW. Model boundary conditions include head-dependent flow to and from streams and head-dependent evapotranspiration. In addition, recharge, pumping, and boundary fluxes from adjacent basins are specified for each stress period. The model was originally calibrated to transient conditions from 1978 through 1997 (referred to as the "original" model period) using semiannual stress periods. The model was not recalibrated for this study. The only changes were the extension of model input files for boundary stresses (recharge, pumping, and evapotranspiration) to cover the period from 1997 through 2008 (referred to as the extended model period). To make the model more compatible with newer stress packages, the model was converted from MODFLOW-96 to MODFLOW-2000.

Documenting Water-Level Changes

The extent and magnitude of water-level changes in the upper Deschutes Basin since 1997 was evaluated by compiling water-level data collected by OWRD between 1997 and 2008. In addition, selected wells that were monitored up to 1997 were revisited and measured. Hydrographs showing changes in water levels with time were updated, and changes in water levels over certain time intervals were plotted on maps. Data were insufficient to constrain the geographic extent of water-level changes, particularly those changes south of Redmond and near Bend. Consequently, it was not feasible to create contour maps showing water-level changes.

Modification of Model Input Files

The existing groundwater model of Gannett and Lite (2004) was used to evaluate the relative influence of factors contributing to water-level changes observed since the mid-1990s. The original model calibration and associated input files were for 1978 through 1997. For this study, model input files for recharge, groundwater pumping, and evapotranspiration were updated to include the period from 1997 through 2008.

Recharge from precipitation, which is affected by climate trends, was calculated for the original model and updated using a daily energy and moisture balance model known as the Deep Percolation Model (DPM) (Bauer and Vaccaro, 1987; Vaccaro, 2007). The DPM calculates recharge using climate observation data (temperature, precipitation, and solar radiation) from weather stations in the basin along with data describing various landscape characteristics. The DPM was used to estimate recharge for the original modeling period by Boyd (1996). To create recharge datasets through 2008, Boyd's recharge model was updated by extending the climate observation data. Other landscape factors were not changed.

Model input files describing the rate and distribution of groundwater pumping for the original model period were updated to cover the extended model period. Estimates of public-supply pumping in the original model were based on data from public water providers in the basin. Public-supply pumping data for the extended model period were obtained directly from water providers as well as water-use reporting records from OWRD. The largest public water suppliers during both the original and extended model periods include the cities of Bend and Redmond, and Avion Water Company.

Groundwater pumping for irrigation was estimated for the original model period (1978–97) using satellite imagery and water-rights mapping from 1994. Satellite images were used to identify crop types growing in areas mapped as irrigated with groundwater. Pumpage estimates were then developed based on water needs of the particular crop types and irrigation efficiency values. Pumping was distributed in the model using surveyed well locations and well log information. The 1994 base-period estimates were distributed to other years using water right priority dates.

Irrigation pumping volumes for the extended model period were developed using water rights data and well log information from OWRD. Consumptive use estimates were based on historical cropping patterns and crop data from the National Agricultural Statistics Service (2007).

Rates of recharge from canal leakage for the original model calibration period were estimated based on diversion and delivery data, seepage and ponding tests, and information on canal-bed geology and geometry (Gannett and others, 2001). Decreases in recharge from canal leakage that occurred during the extended model period were calculated based on a compilation of estimated decreases in canal leakage for pipe, lined, or abandoned canals provided by OWRD (Jonathan La Marche, written commun., 2009). The OWRD compilation included a geographic information system (GIS) map that showed the affected sections of canal, and the year that canal modification construction began and ended. For this study, decreases in canal leakage were assumed to commence the year canal modifications were completed. Decreases in canal leakage were provided for 225 separate canal or lateral reaches. These decreases in leakage were summed for each model grid cell for the appropriate years, and then subtracted from the leakage estimates used during the original model period.

Estimates of groundwater recharge from on-farm losses of irrigation water for the original model period were based on literature values coupled with knowledge of irrigation application methods in various areas (Gannett and others, 2001). On-farm loss refers to applied irrigation water that percolates beneath the rooting depth of plants that becomes groundwater recharge. On-farm losses were calculated as the difference between applied irrigation water (after factoring in irrigation efficiency) and consumptive use of the associated crops. On-farm losses were decreased in the extended model period based on annual (2001–08) estimates provided by OWRD of land area not irrigated to decrease surface-water diversions (Kyle Gorman, OWRD, written commun., 2009). In most cases, decreases in on-farm losses were distributed uniformly over the associated irrigation district because there was no readily available information about specific locations. Decreases in on-farm losses in Central Oregon Irrigation District area were focused around Redmond, per information from OWRD. Decreases in recharge from on-farm losses during the extended model period were calculated by dividing the non-irrigated acreage by the total district acreage, and decreasing the historical on-farm loss recharge by that proportion.

Groundwater Flow Simulation Analysis

To validate the ability of the model as originally calibrated to simulate conditions from 1997 to 2008, the model was run with all appropriate stresses updated (as described in section, "Modification of Model Input Files") and simulated heads and flows were compared to measured equivalents during that period. The model fit to measured heads and flows during the extended model period was similar to the fit during the original model period, and the model was deemed suitable for examining conditions during the extended model period. To evaluate the relative contributions of climate variations, groundwater pumping, and piping and lining of irrigation canals to measured water-level changes, the model was run with all input files updated to 2008 (the base run), and then the model was run holding pumping and/or canal recharge rates at 1994 levels. We chose 1994 as the departure point for comparison because it was near the low point of the last drought cycle in the original model period. It is also the year where base estimates for pumping and canal leakage were determined for the original model period. The influence of individual stresses (climate variations, increased groundwater pumping, and decreased recharge due to canal lining) were evaluated by comparing simulation results of the base run with results from runs with individual stresses held at 1994 levels.

Changes in Hydrologic Conditions

Groundwater-Level Changes

Groundwater-level changes were evaluated using periodic (generally quarterly) measurements made by OWRD in approximately 25 wells from the mid-1990s through 2008. The locations of selected observation wells are shown in figure 4. Water levels in five of the monitored wells are of no value in evaluating climate or pumping effects because they are influenced primarily by the stages of nearby streams or lakes that are artificially manipulated. Water levels in most of the remaining wells reflect varying influences of pumping and climate variations.

Water-level changes that occurred over three time periods (1996–2000, 2000–04, and 2004–08) throughout the upper Deschutes Basin are shown in figure 5. The time periods approximately correspond to predominantly wet or dry periods. Water levels do not change uniformly over the entire basin (fig. 5), and different areas show different magnitudes and directions of water-level change each time period. The central part of the upper Deschutes Basin, specifically the area from Cline Buttes east to the community of Powell Butte,

experienced consistent water-level declines between 1996 and 2008. Well D in figure 2 is an example of a well exhibiting a consistent water-level decline.

Wells in and adjacent to the Cascade Range showed water-level fluctuations generally following climate variations (fig. 2, well A). Water levels generally rose during 1996–2000 in response to wet conditions, with many wells showing rises of more than 10 ft. During the dry period from 2000 through 2004, water levels in the Cascade Range declined by amounts similar to the earlier water-level rise. Between 2004 and 2008, water levels showed moderate rises in those wells that were closest to the Cascades and not influenced by pumping or stream stage.

Wells in the La Pine subbasin south of Bend also tend to respond to climate cycles, and show no evidence of discernible pumping-related trends due to distance from large pumping centers. During the three climate periods shown in figure 5, shallow wells in the La Pine subbasin exhibited decadal patterns of fluctuation similar to those in the Cascades. Decadal climate fluctuations are relatively small and are of similar magnitude to seasonal variations (fig. 6). The more subdued fluctuations likely are a reflection of the low overall recharge rates in the area and distance from the Cascade Range.

Groundwater levels appear to have declined almost continuously in the area extending from Cline Buttes east to Powell Buttes, and from Cline Buttes north toward Lower Bridge. As shown by representative hydrographs (figs. 7 and 8), water-level trends in this area are characterized by subtle climate fluctuations superimposed on a dominant post-1990 downward trend. The climate influences are insufficient to overcome the downward trend, and the wet conditions in 1996 to 2000 that resulted in substantial water-level rises in wells to the west only resulted in a lessening of the decline rate in wells in this area.

Wells that have been monitored in the northernmost part of the upper Deschutes Basin, northeast of the Crooked River in Crook and Jefferson Counties, generally show moderate water-level fluctuations that tend to follow climate fluctuations. As recently as the late 1990s, water levels were still rising in some wells in Jefferson County, presumably in response to construction of Round Butte Dam and the creation of Lake Billy Chinook (Gannett and others, 2001).

The continuous water-level declines in parts of the upper Deschutes Basin may not solely be the result of pumping stresses. The declines are likely influenced by the general drying trend in the basin over the past several decades. It is possible that the effects of decadal wet and dry cycles are largely diffused and attenuated due to distance from the Cascade Range (the principal recharge area) and that the wells are responding to the longer term climate pattern.

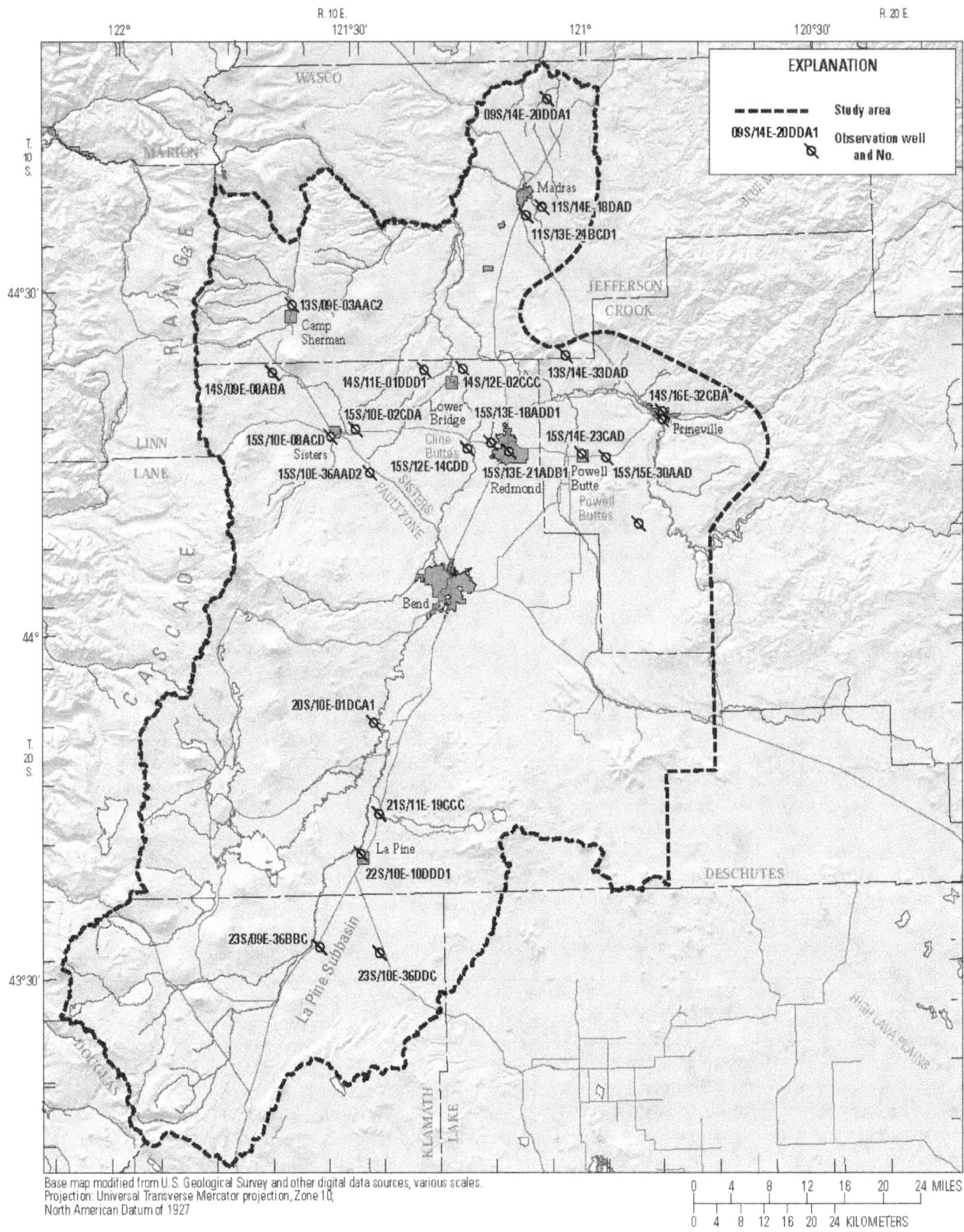

Figure 4. Locations of selected observation wells in the upper Deschutes Basin, central Oregon.

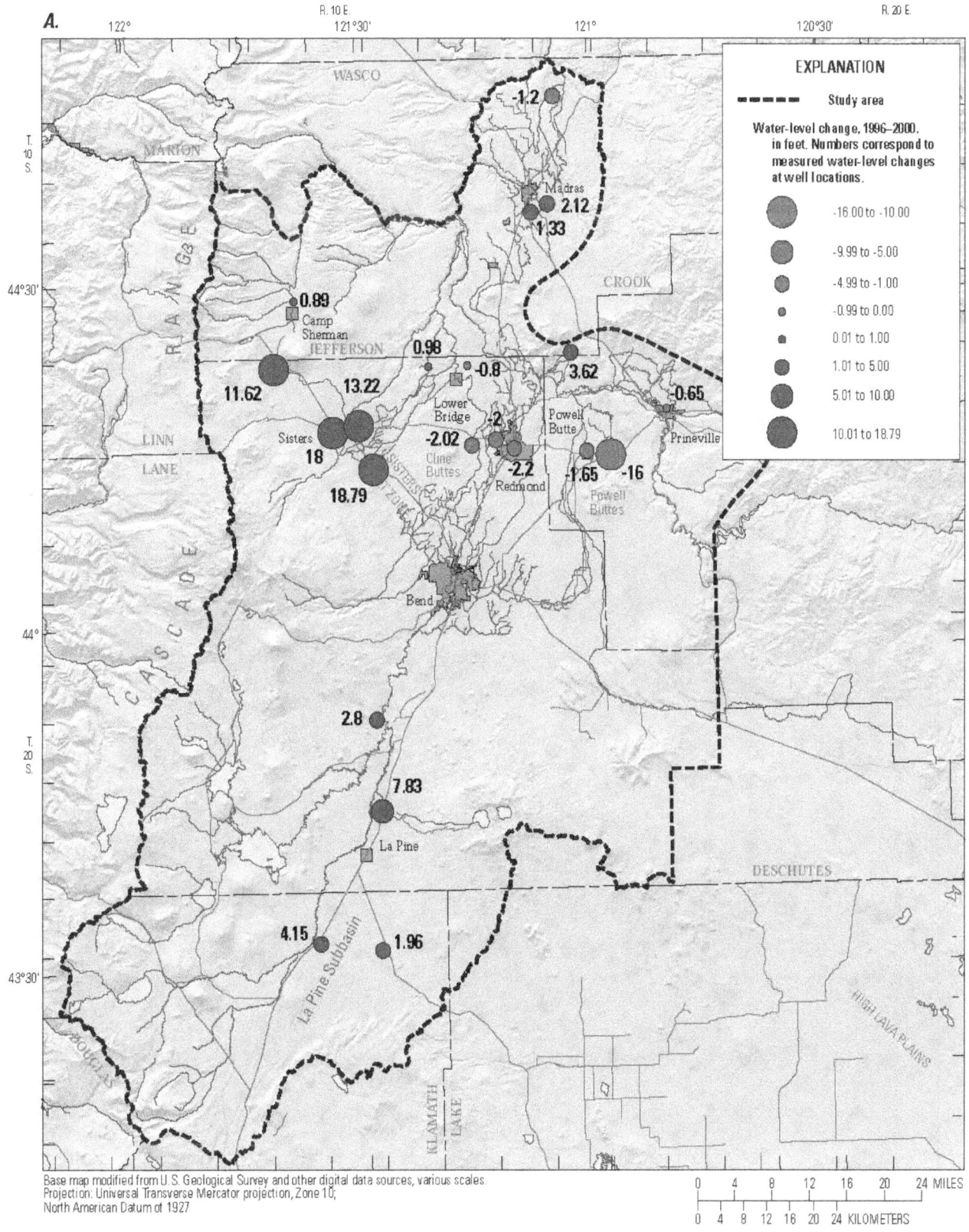

Figure 5. Changes in measured springtime high-water levels in selected observation wells in the upper Deschutes Basin, central Oregon. (*A*), 1996–2000; (*B*), 2000–04; (*C*), 2004–08.

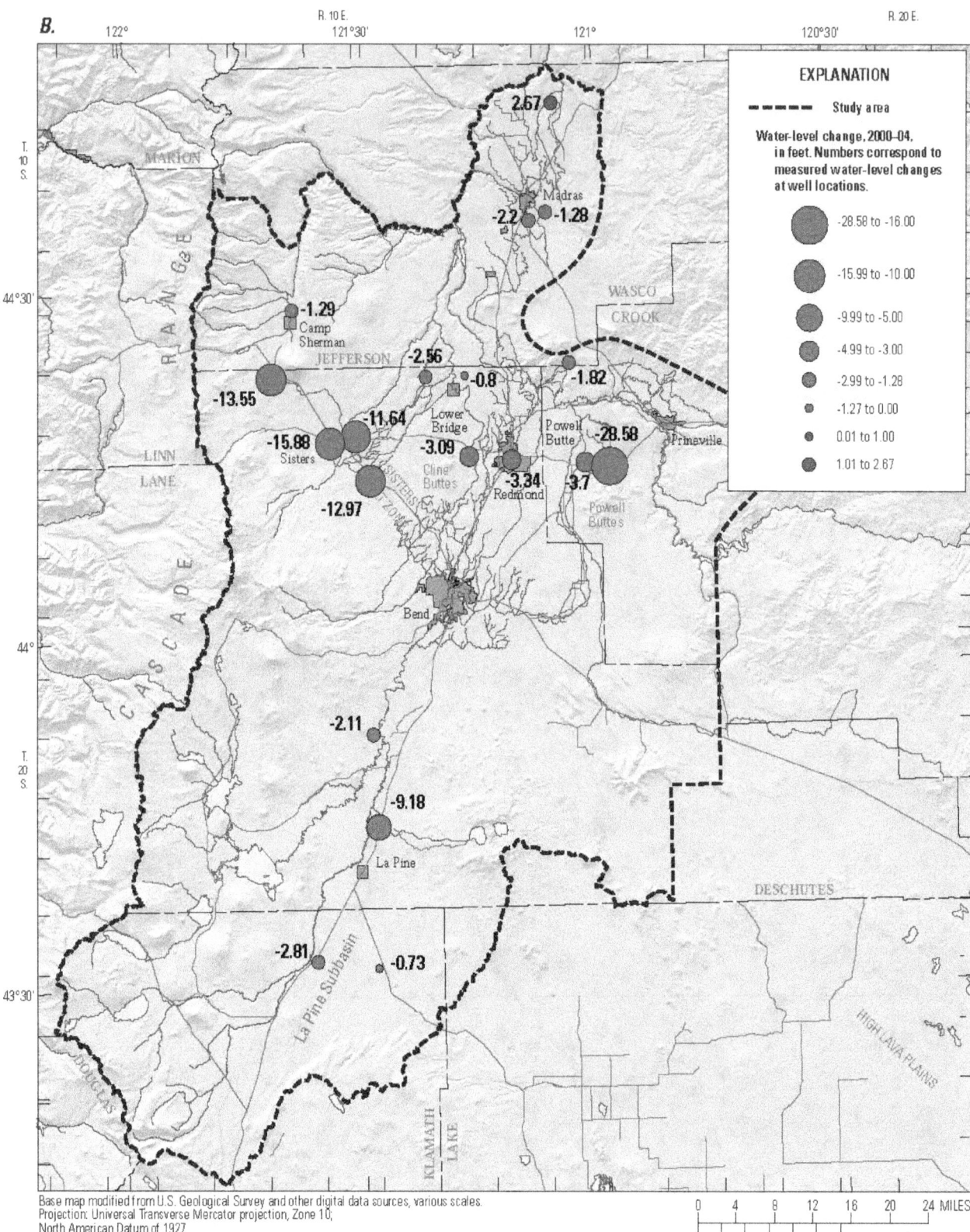

Figure 5.—Continued

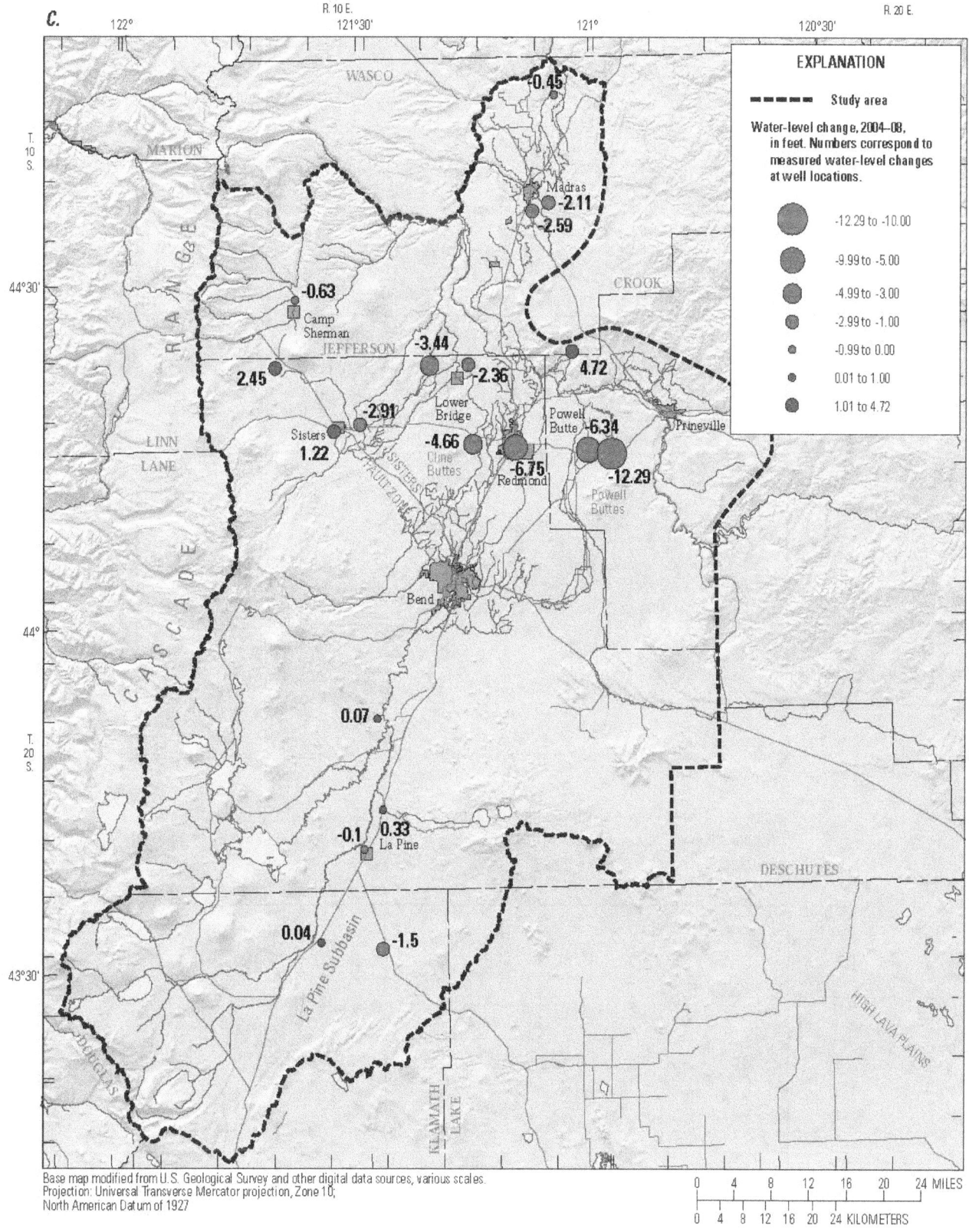

Figure 5.—Continued

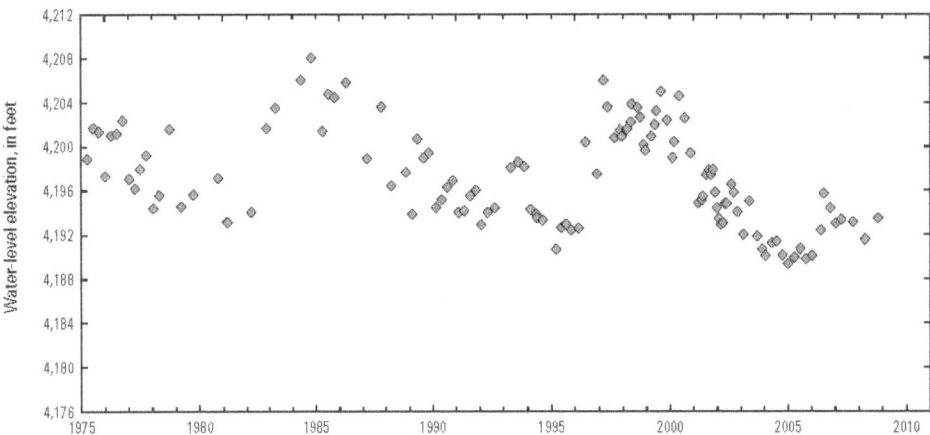

Figure 6. Water-level measurements in observation well 21S/11E-19CCC in the La Pine subbasin, central Oregon, showing seasonal variations as well as the effects of decadal climate fluctuations. Location of observation well is shown in figure 4.

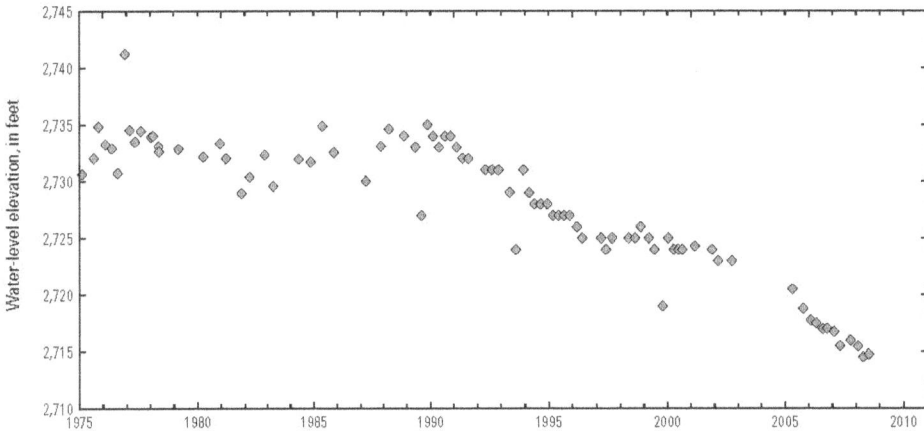

Figure 7. Water-level measurements in observation well 15S/13E-18ADD1 near Redmond, Oregon, showing continuous water-level declines since the early 1990s. Location of observation well is shown in figure 4.

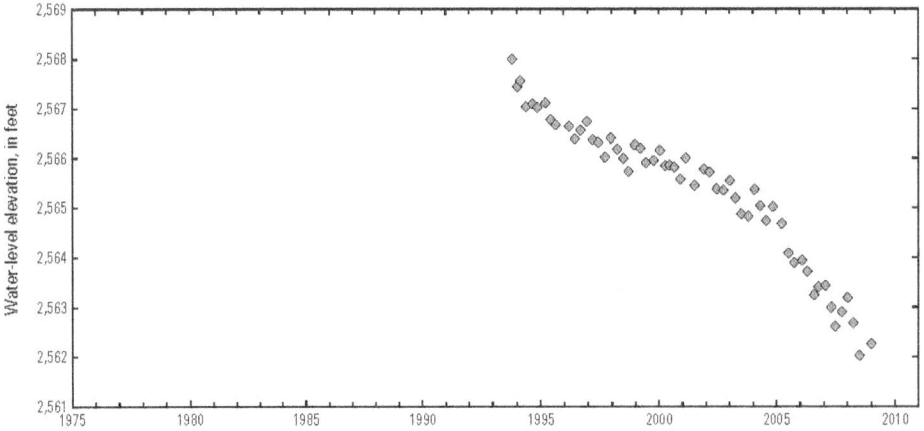

Figure 8. Water-level measurements in observation well 14S/12E-02CCC near Lower Bridge, northeast of Sisters, Oregon, showing continuous water-level declines since the early 1990s. Location of observation well is shown in figure 4.

Changes in Groundwater Recharge from Precipitation

Deep percolation model results show that groundwater recharge in the upper Deschutes Basin has decreased in recent decades due to the drying climate trend (fig. 9). Average annual recharge rates for 1979–88, 1989–98, and 1999–2008, were 3.2, 2.8, and 2.4 million acre-ft/yr, respectively. Groundwater recharge decreased about 25 percent between the 1979–88 and 1999–2008 periods, at a long-term average rate of about 33,500 acre-ft/yr. The decrease in groundwater recharge is manifested as a decrease in discharge of most spring-fed streams. For example, mean annual discharge of Fall River, a spring-fed stream (fig. 3), has decreased from 135 ft³/s (1978–88) to 118 ft³/s (1999–2008), a decrease of about 13 percent. Decreases in mean annual discharge of Fall River between the 1970s (150 ft³/s) and 2000s (110 ft³/s) is even larger, approaching 27 percent. Decreases in recharge and discharge in spring-fed streams are consistent with decreased discharge of groundwater-dominated streams over the past 50 years elsewhere in the Cascade Range documented by Mayer and Naman (2011).

Decreases in Recharge from Lining and Piping Irrigation Canals

Information from OWRD (J.L. La Marche, written commun., 2009) indicates there was substantial lining and piping of irrigation canals in the upper Deschutes Basin between 1994 and 2008. By 2008, canal leakage was reduced by approximately 58,000 acre-ft/yr, a decrease of 16 percent compared to the 356,600 acre-ft/yr in 1994 as estimated by Gannett and others (2001). The decrease in annual groundwater recharge as a result of lining and piping of irrigation canals between 1994 and 2008 is shown in figure 10.

Decreases in Recharge from On-Farm Losses

The OWRD provided a compilation of acreage for which surface-water rights were not exercised to reduce diversions and allow water to remain instream. Reductions in irrigated land area result in decreases in recharge from on-farm losses. Acreage amounts were provided for each irrigation district, as well as non-district areas from 2001 to 2008. The area of idled land varied year to year, generally increasing from 2001 to 2008 with annual totals ranging from about 2,000 to 8,000 acres (J.L. La Marche, OWRD, written commun., 2012). The proportion from non-district areas averaged about 12 percent of the total. Decreases in groundwater recharge resulting from idling land from 2001 to 2008 ranged from approximately 250 to 1,000 acre-feet/yr. This is a small fraction of the approximately 49,000 acre-ft of annual recharge from on-farm losses (Gannett and others, 2001).

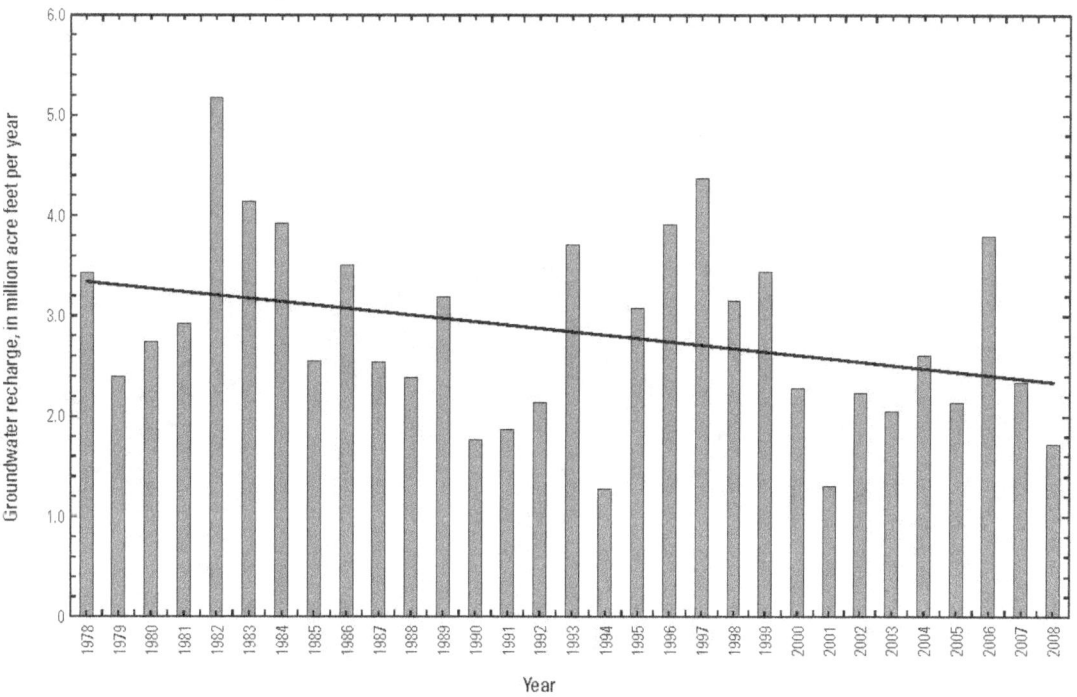

Figure 9.   Estimated total annual groundwater recharge in the upper Deschutes Basin, central Oregon, 1978–2008. Line shows linear trend with a slope of about 33,500 acre-ft/yr.

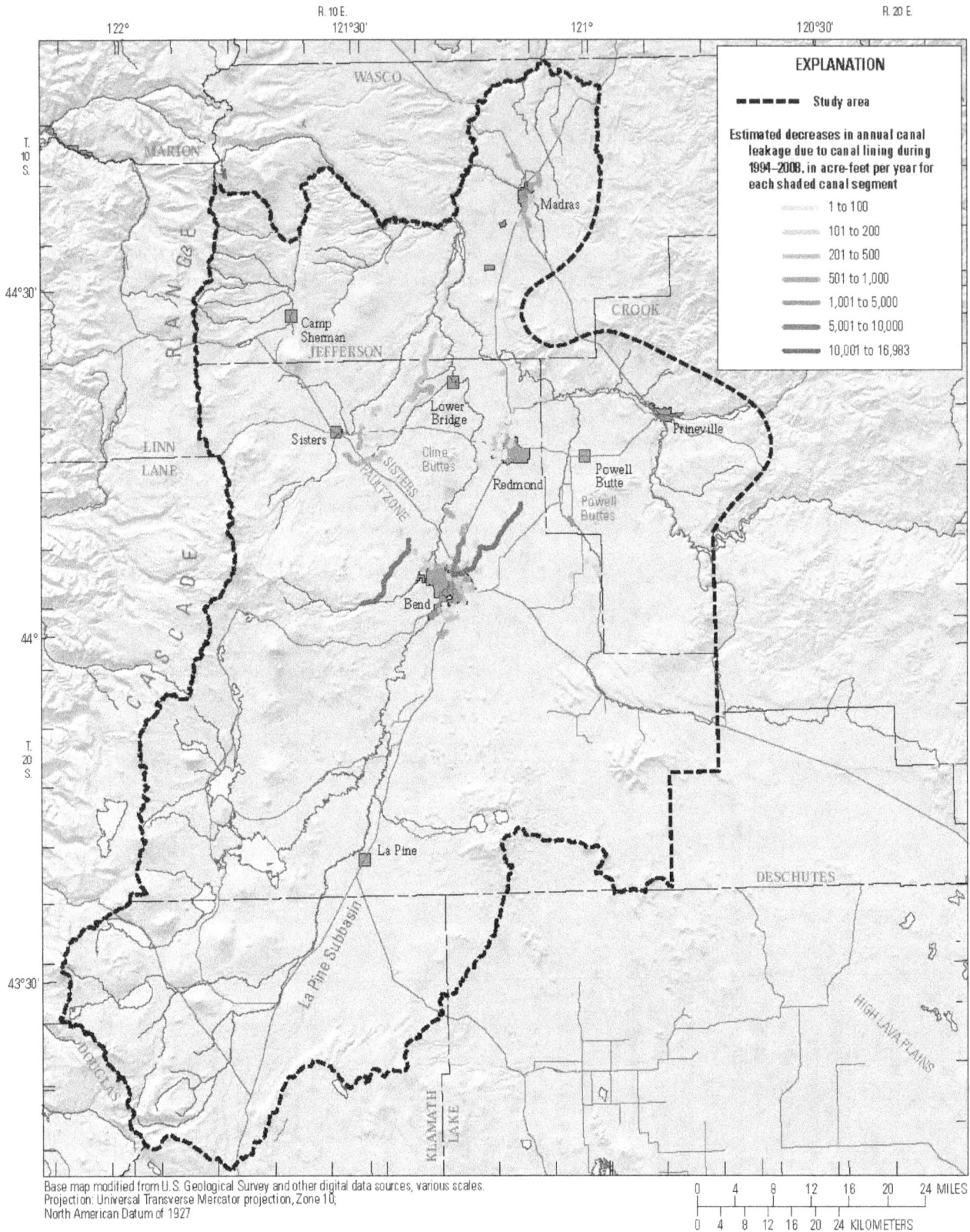

Figure 10.   Spatial distribution and magnitude of estimated decreases in annual canal leakage due to lining or piping of irrigation canals between 1994 and 2008 in the upper Deschutes Basin, central Oregon. Data from the Oregon Water Resources Department.

Increases in Pumping

Pumping volumes for municipal and irrigation uses increased about 68 percent between 1994 and 2008. Pumping for irrigation increased from about 17,500 to about 25,000 acre-ft/yr (fig. 11). The different methods of estimating pumpage used for the original model period and the extended model period resulted in slightly different estimates for 1994 (15,000 and 17,500 acre-ft, respectively). For base case simulations used in this analysis, the new estimates were used from 1994 through 2008.

Public-supply pumping increased from 13,400 acre-ft/yr in 1994 to close to 26,800 acre-ft/yr in 2007 (fig. 12).

Reported public-supply pumping decreased to 23,400 in 2008, probably due to normal year-to-year variations in use. Increases in pumping by the Cities of Bend and Redmond, as well as the Avion Water Company, accounted for most of the increase in public-supply use.

Although groundwater pumping has increased, the general distribution is unchanged (fig. 13), as it largely follows established land-use patterns. Public-supply pumping continues to be concentrated in the areas of Bend and Redmond, while irrigation pumping continues to dominate near Sisters, the Lower Bridge area, and north of Powell Buttes.

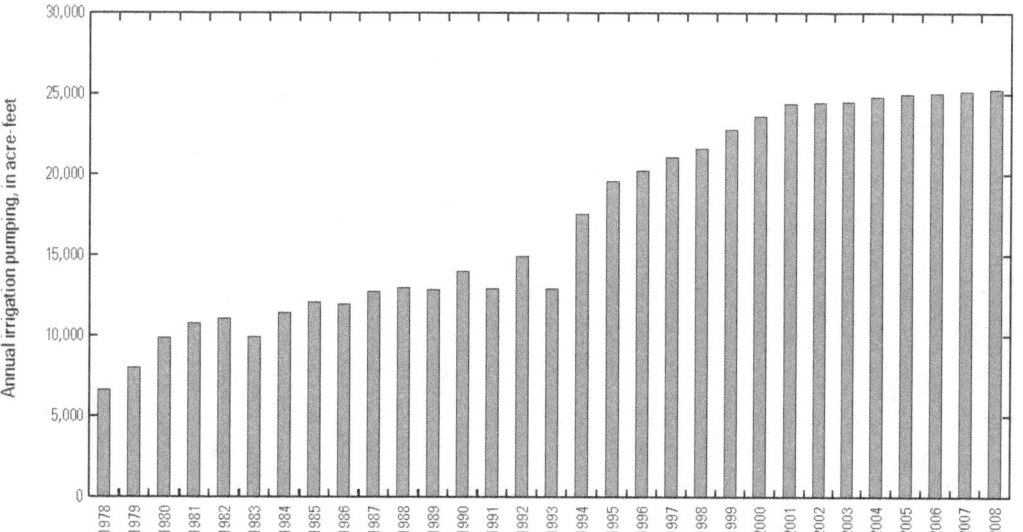

Figure 11. Estimated annual groundwater pumping for irrigation in the upper Deschutes Basin, central Oregon, 1978–2008.

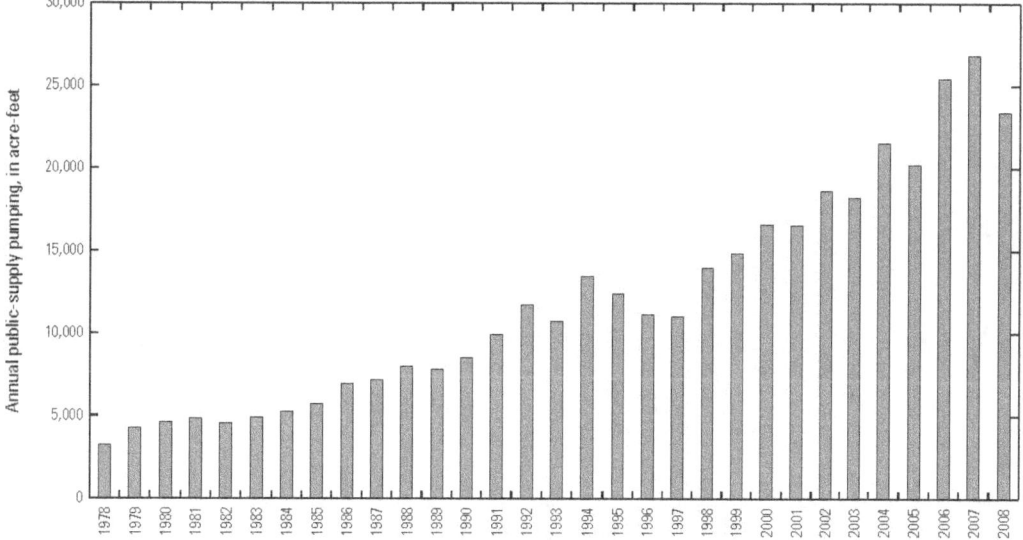

Figure 12. Estimated annual public-supply pumping in the upper Deschutes Basin, central Oregon,1978–2008.

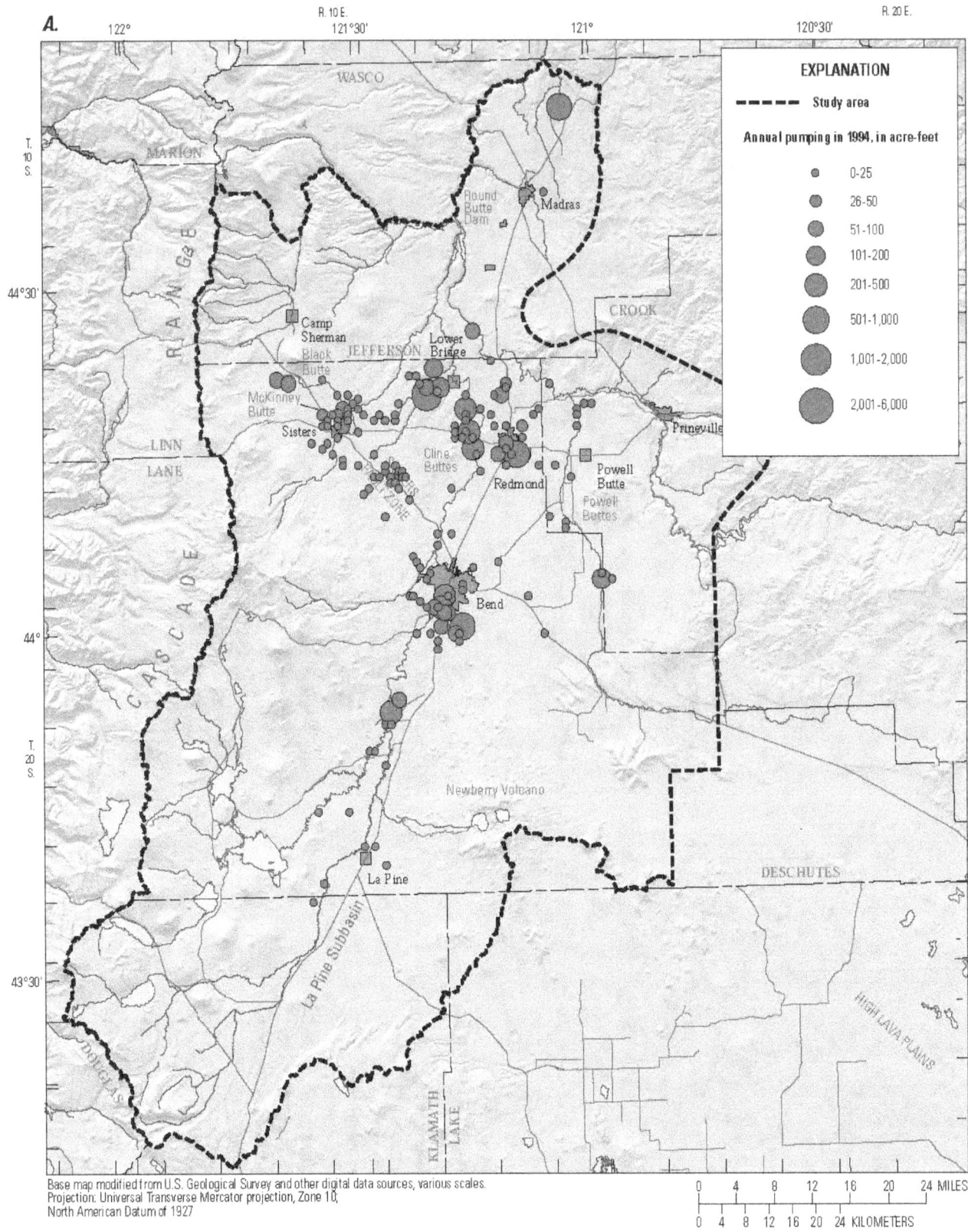

Figure 13.   Distribution of groundwater pumping in the upper Deschutes Basin, central Oregon. (*A*), 1994; (*B*), 2008; (*C*), change from 1994 to 2008 (negative values indicate decreases).

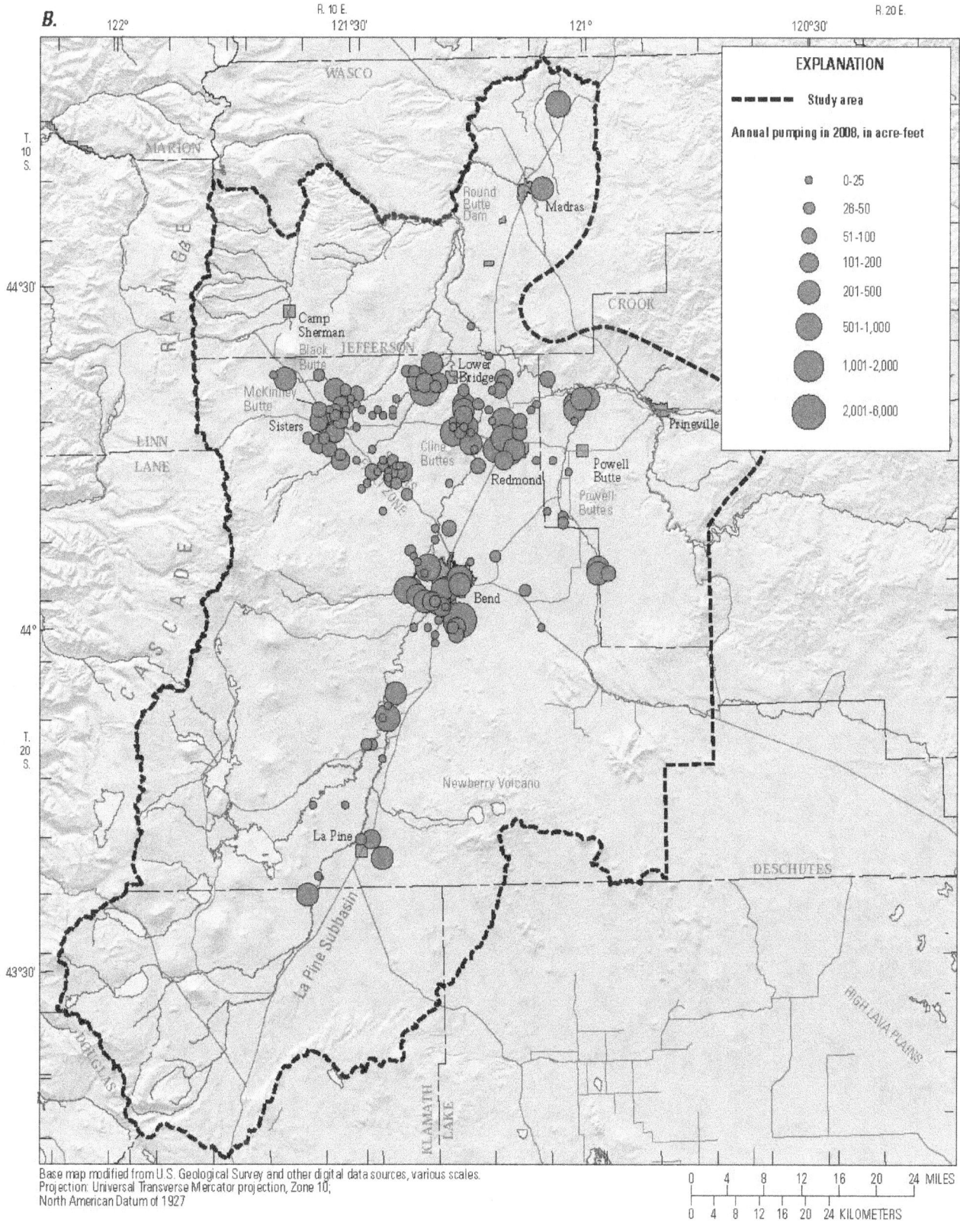

B.

EXPLANATION

- - - - Study area

Annual pumping in 2008, in acre-feet

- 0-25
- 26-50
- 51-100
- 101-200
- 201-500
- 501-1,000
- 1,001-2,000
- 2,001-6,000

Base map modified from U.S. Geological Survey and other digital data sources, various scales.
Projection: Universal Transverse Mercator projection, Zone 10;
North American Datum of 1927

0   4   8   12   16   20   24 MILES

0   4   8   12   16   20   24 KILOMETERS

Figure 13.—Continued

Figure 13.—Continued

Simulation of 1997–2008 Groundwater Conditions

The original model calibration period, which spans the period from 1978 to 1997, included both wet and dry climatic cycles. The varying conditions were reflected in the hydrologic measurements used to calibrate the model. The original model fit was evaluated by comparing measured heads and discharges to their simulated equivalents (Gannett and Lite, 2004). Fitted error statistics (Hill and Tiedeman, 2007) indicate that 68 percent of simulated head values are within about 76 ft of the measured values, and 95 percent of simulated head values are within 152 ft of the measured values. This fit is considered good, given that measured heads vary more than 4,500 ft over the 4,000 mi² model domain. Although the absolute values of measured and simulated heads commonly vary tens of feet, water-level fluctuations, or changes, match much more closely. Gannett and Lite (2004) show that the model does a good job of capturing the temporal variation in hydraulic heads and groundwater discharge caused by climate and other external stresses.

Model runs incorporating the extended model period show that the model does a good job of simulating measured fluctuations in hydraulic heads and groundwater discharge through 2008. Figure 14 shows simulated groundwater discharge to Fall River and Odell Creek along with measured streamflow. Fall River is entirely spring-fed, so the streamflow is directly comparable to the simulated groundwater discharge. Temporal variations in simulated groundwater discharge to Fall River match those of the measured streamflow reasonably well for the original and extended model periods, although absolute discharge volumes are high by 50–75 ft³/s (fig. 14A). Odell Creek is not entirely groundwater fed, and the streamflow hydrograph shows peaks resulting from storms and snowmelt events superimposed on a relatively robust baseflow of about 50–100 ft³/s. Comparison of simulated groundwater discharge and measured streamflow at Odell Creek shows that the model does a good job simulating the baseflow component of flow in both terms of temporal variations and magnitude through the original and extended model periods.

The model did a generally good job of simulating water-level changes during the original calibration period. In the central part of the upper Deschutes Basin where water-level declines are largest, the simulated water-level changes match observations quite well, although there is an offset in absolute head values of about 25 ft (fig. 15). Comparison of simulated and measured water-level changes from 1997 to 2008

indicates that the model also performs well in the extended period. Because the focus of this study is to understand the relative magnitude of water-level changes attributable to various stresses, it is most important that the model do an adequate job of simulating the measured temporal variations, and matching absolute heads is less important. Because the model matches measured temporal variations in water levels during the original and extended model periods reasonably well, particularly in the central part of the upper Deschutes Basin, it is considered an appropriate tool for the purposes of this study.

Relative Effects of Climate, Pumping, and Canal Lining

The regional groundwater model was used to determine the relative influence of climate variations, decreased canal leakage, and increased pumping on measured water-level changes. The effects of each of these stresses were isolated by running three simulations, each one including a different combination of the stresses. In the first simulation, climate variations, increased pumping, and decreased canal leakage up to 2008 were all included. This simulation, referred to as the base run, reflects actual conditions and was used to compare simulated and measured water levels. For the second simulation, the recent increase in groundwater pumping was removed by holding post-1994 pumping rates at the 1994 level. Results of this model run are compared to the base run to evaluate the relative influence of the increase in pumping. For the third simulation, both pumping and canal-leakage rates were held constant after 1994. This shows the influence of climate variations alone, and also allows evaluation of the relative influence of decreased canal leakage. The effects of changes in on-farm losses were sufficiently small that they are not considered in the analysis (they were, however, simulated in all model runs). The simulated water-level changes were evaluated at locations where water levels have been monitored since the mid-1990s (fig. 5).

Water-level changes in some parts of the upper Deschutes Basin, such as the La Pine subbasin and upland areas, are due to climate influences and are largely unaffected by pumping and canal lining. Simulated water levels in the more developed central part of the upper Deschutes Basin (the area encompassing Sisters, Bend, Redmond, and Powell Butte) show the effects of increased pumping and decreased recharge due to canal lining in addition to climate variations.

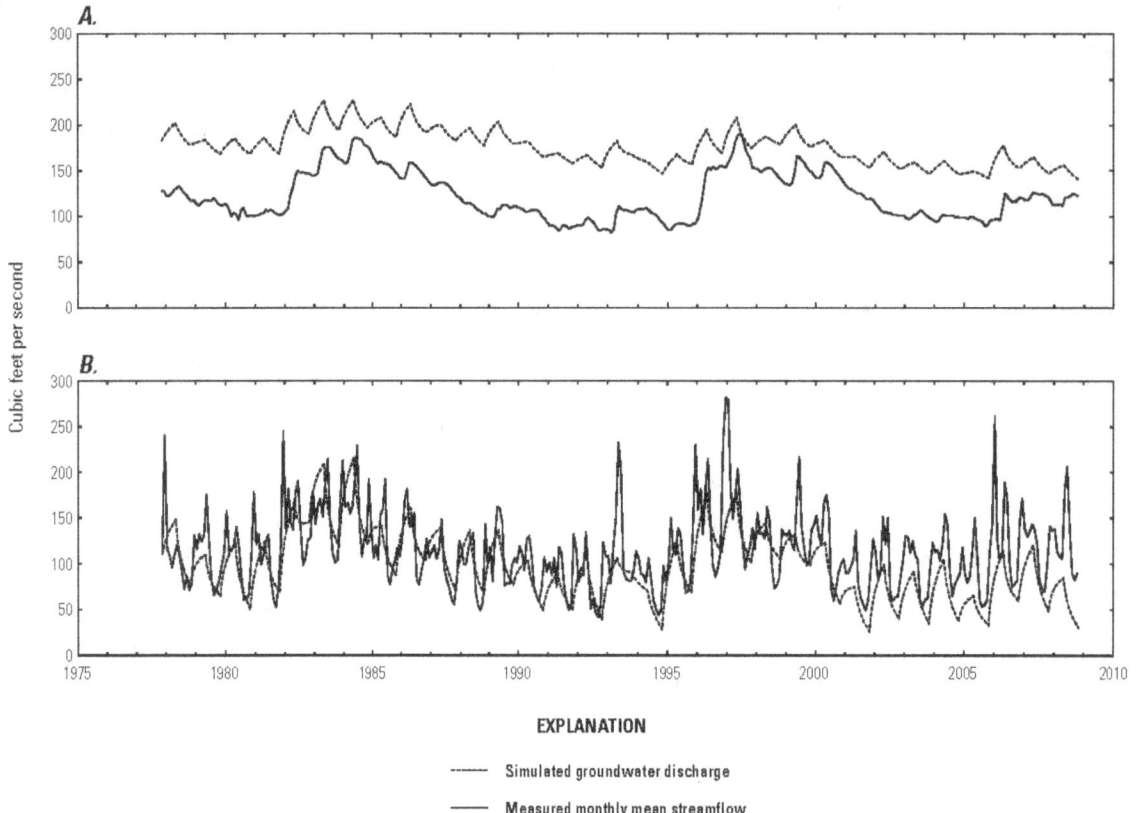

Figure 14. Measured streamflow and simulated groundwater discharge to *A*, Fall River (USGS gage 14057500) and *B*, Odell Creek (USGS gage 14055600), upper Deschutes Basin, central Oregon. Locations of streamflow-gaging stations are shown in figure 1.

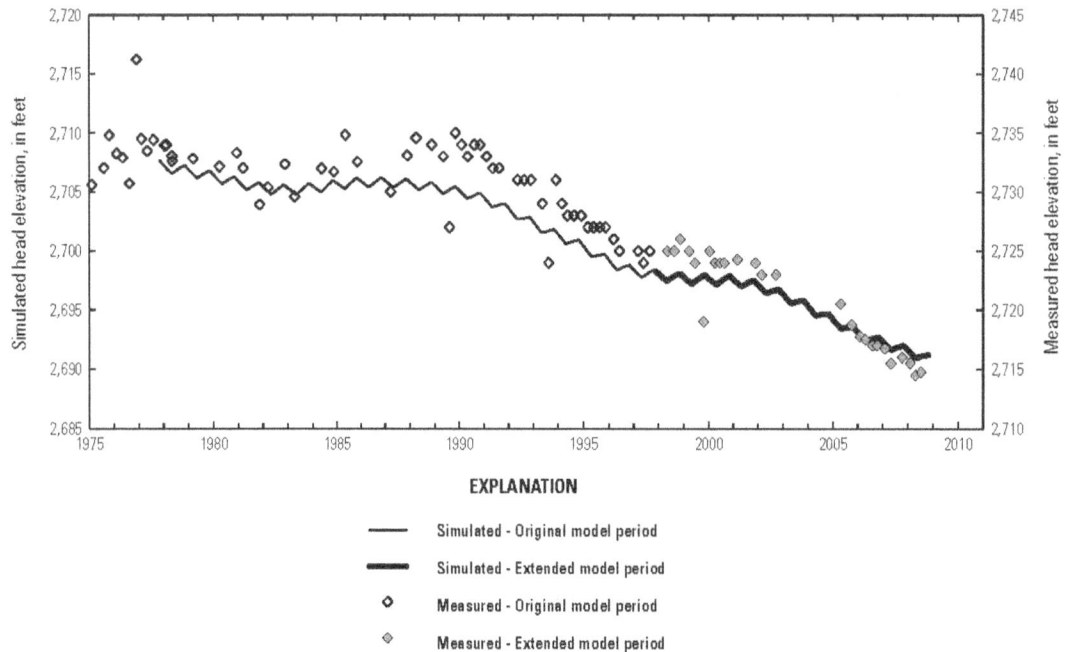

Figure 15. Simulated and measured head elevation in observation well 15S/13E-18ADD1 near Redmond, Oregon. Location of observation well is shown in figure 4.

La Pine Subbasin and Cascade Range

The effects of increased pumping and decreased canal leakage range from small to undetectable in the La Pine subbasin and along the western margin of the upper Deschutes Basin in and near the Cascade Range. In the La Pine subbasin, the effects of decreased canal leakage are too small to be seen on graphs comparing the relative influence of stresses and are not included in the associated figures. The effects of pumping increases since 1994 cannot be seen in shallow parts of the system in the La Pine subbasin, probably due to the relatively small increases in local pumping (fig. 16). In deep wells, post-1994 pumping increases account for about 0.5 ft of the roughly 7-ft net decline in water levels measured since the mid-1990s (fig. 17). The pumping influence seen in deep

zones in the La Pine subbasin may be diminished in shallower depths because of the presence of a thick sequence of saturated fine-grained deposits in the area.

Water levels in the few monitoring wells in and adjacent to the Cascade Range also appear to be minimally affected by post-1994 canal lining and increases in pumping. Pumping influences cannot be discerned on a graph showing simulated water levels in a shallow well in the Camp Sherman area (fig. 18), as the plots with and without post-1994 pumping essentially overlie one another. A post-1994 water-level decline of several tenths of a foot due to pumping was simulated in a well south of Black Butte (fig. 19). Canal lining influences are not discernible on simulated hydrographs for either well monitored in the Cascade Range.

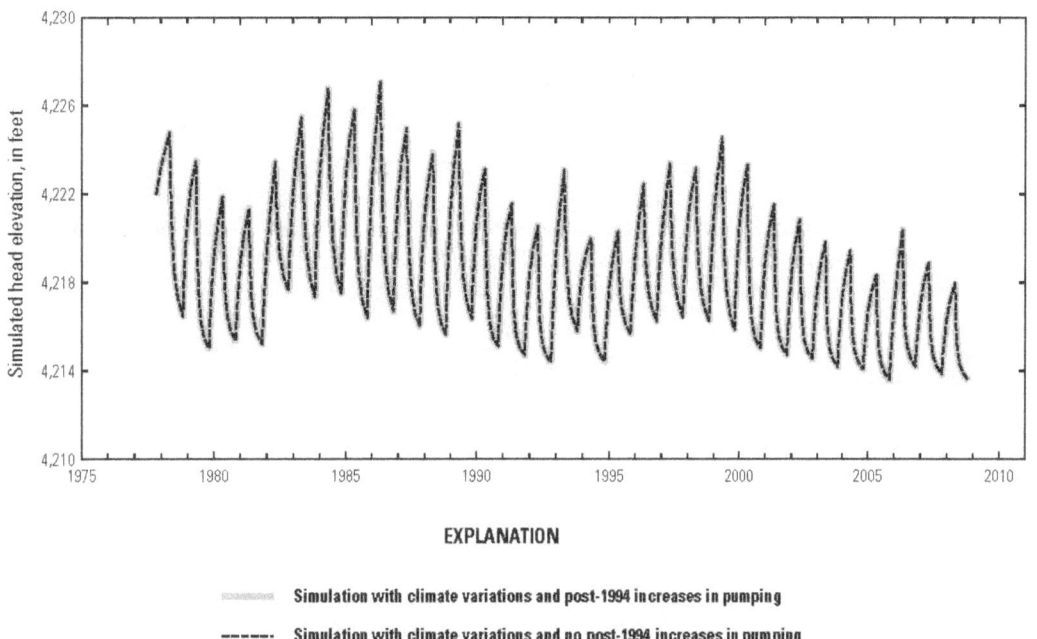

EXPLANATION

Simulation with climate variations and post-1994 increases in pumping

------- Simulation with climate variations and no post-1994 increases in pumping

Figure 16.   Simulated head elevations in observation well 21S/11E-19CCC, a 100-foot deep well in the La Pine subbasin, central Oregon. Lines showing simulated head elevations with and without post-1994 pumping increases are coincident on the graph, indicating very limited impact from post-1994 pumping increases. Effects of post-1994 canal lining are too small to show at the scale of this graph. Location of observation well is shown in figure 4.

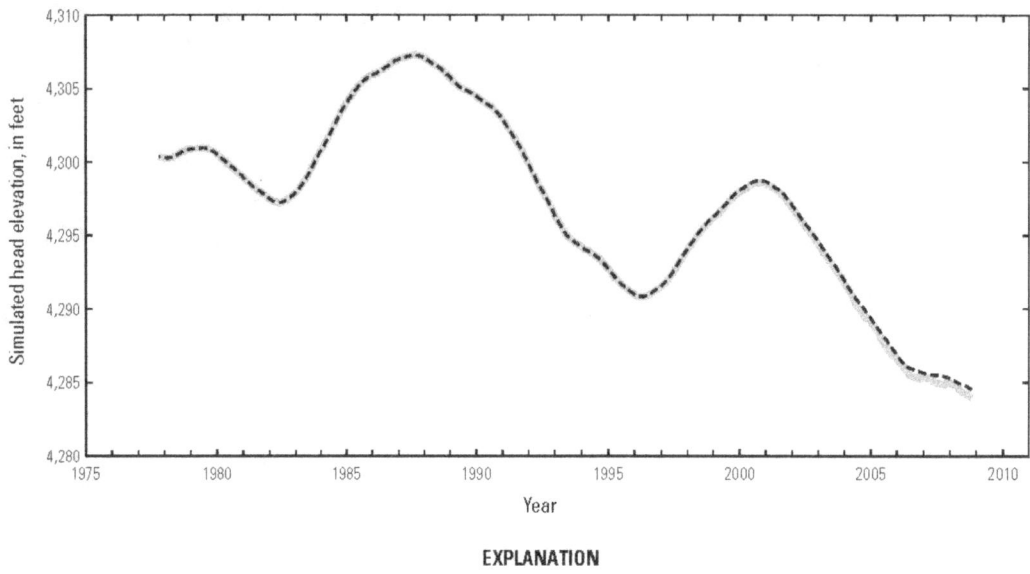

Figure 17.   Simulated head elevations in observation well 22S/10E-10DDD01, a 1,458-foot deep well near La Pine, central Oregon. Comparison of simulated heads with and without the post-1994 groundwater pumping increases indicates that post-1994 growth in pumping accounts for about 0.5 foot of water-level decline as of 2008. Effects of post-1994 canal lining are too small to show at the scale of this graph. Location of observation well is shown in figure 4.

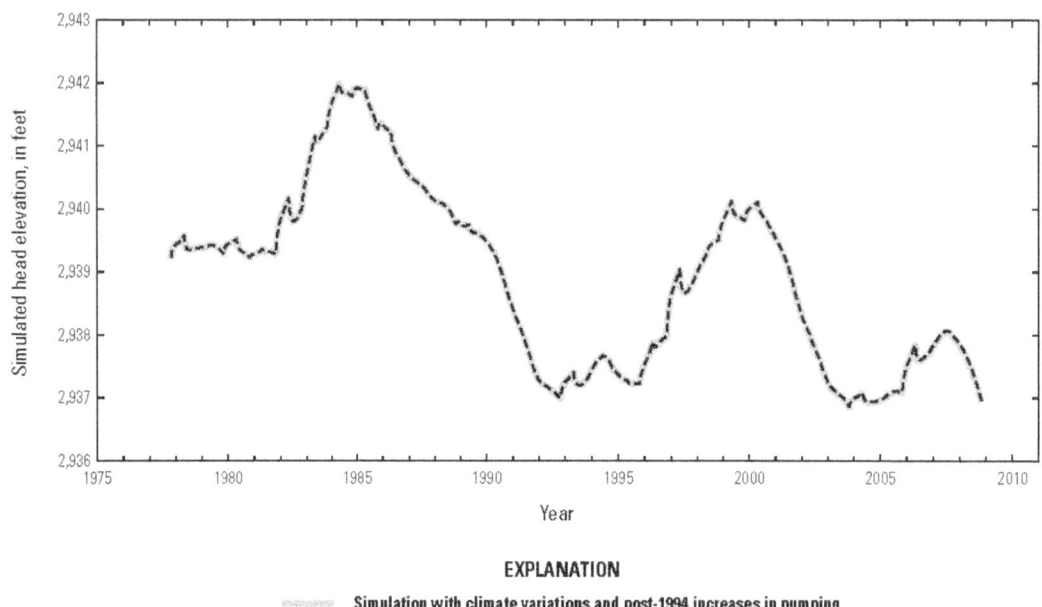

Figure 18.   Simulated head elevations in observation well 13S/09E-03AAC02 near Camp Sherman, Oregon. Lines showing simulated heads with and without post-1994 groundwater pumping increases cannot be discriminated on the graph, indicating very limited impact from post-1994 pumping increases. Effects of post-1994 canal lining are too small to show at the scale of this graph. Location of observation well is shown in figure 4.

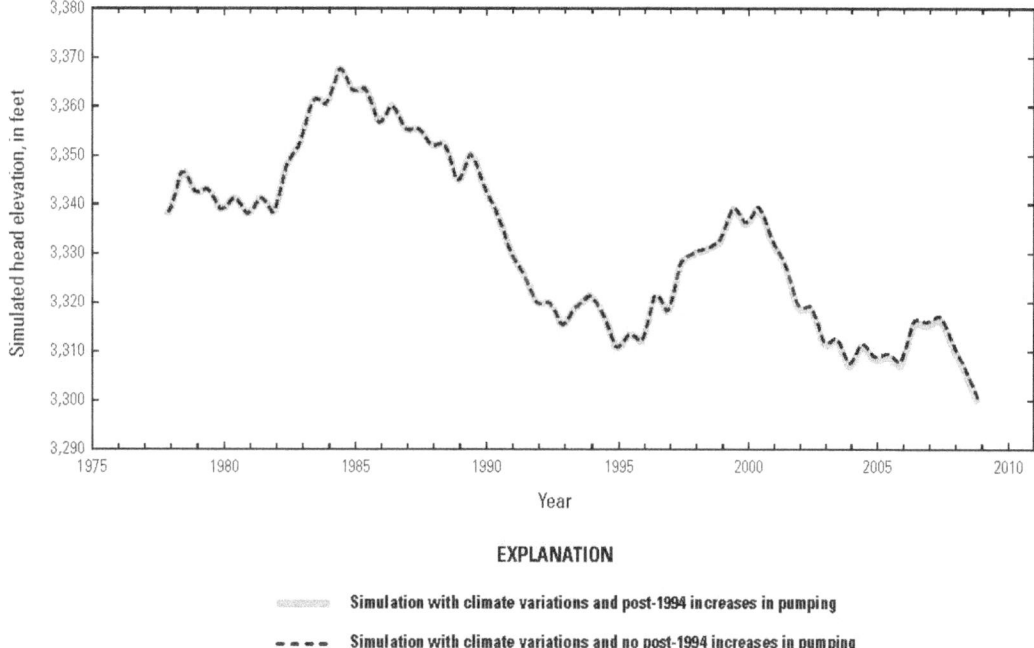

EXPLANATION

⬛⬛⬛   **Simulation with climate variations and post-1994 increases in pumping**

▬ ▬ ▬ ▬   **Simulation with climate variations and no post-1994 increases in pumping**

Figure 19. Simulated head elevations in observation well 14S/09E-08ABA south of Black Butte, Oregon. Comparison of simulated heads with and without post-1994 groundwater pumping increases indicates that post-1994 growth in pumping accounts for water-level declines of several tenths of a foot. Effects of post-1994 canal lining are too small to show at the scale of this graph. Location of observation well is shown in figure 4.

Sisters Area

Water-level trends in the Sisters area differ on either side of McKinney Butte. McKinney Butte lies along the Sisters fault zone, which demarks a transition zone east of which water-level fluctuations observed in the Cascade Range become diffused and attenuated. West of the McKinney Butte, measured water levels have risen about 10 ft since 2005 in a manner similar to that observed in the Cascade Range. Such a post-2005 rise in water levels generally is not observed east of McKinney Butte.

Simulations show that about 80 percent of the roughly 22-foot water-level declines in the western part of the Sisters area since the peak of the most recent wet period (about 2000) are due to climate, and that pumping and canal lining are responsible for approximately 13 and 7 percent of declines, respectively (fig. 20). Of the 20- to 25-ft decline in water levels observed just east of the Sisters area since 2000, about 65–70 percent can be attributed to climate. The remaining 30–35 percent (about 6–8 ft) can be attributed to increased pumping and canal lining in nearly equal proportions (fig. 21).

Lower Bridge Area

Water levels in the two wells monitored in the Lower Bridge area northwest of Sisters have declined about 5–6 ft since the mid-1990s (figs. 22 and 23). The decline has been more or less continuous except during the wet period between the mid-1990s and 2000, a period during which water levels rose very slightly (fig. 22), or the decline rate flattened out (fig. 23). The general lack of a significant water-level rise in the Lower Bridge area in response to wet conditions during the late 1990s is typical of wells in the central part of the upper Deschutes Basin around Redmond. It is probable that recharge pulse during this relatively short wet period was largely attenuated by diffusion as it moved west from the Cascade Range and was insufficient to overcome the longer term drying trend apparent in the central part of the basin. Simulations in the Lower Bridge area show that about 60–70 percent of the water-level decline measured since the mid-1990s can be attributed to climate, about 20–30 percent can be attributed to increases in groundwater pumping, and about 10 percent is due to canal lining.

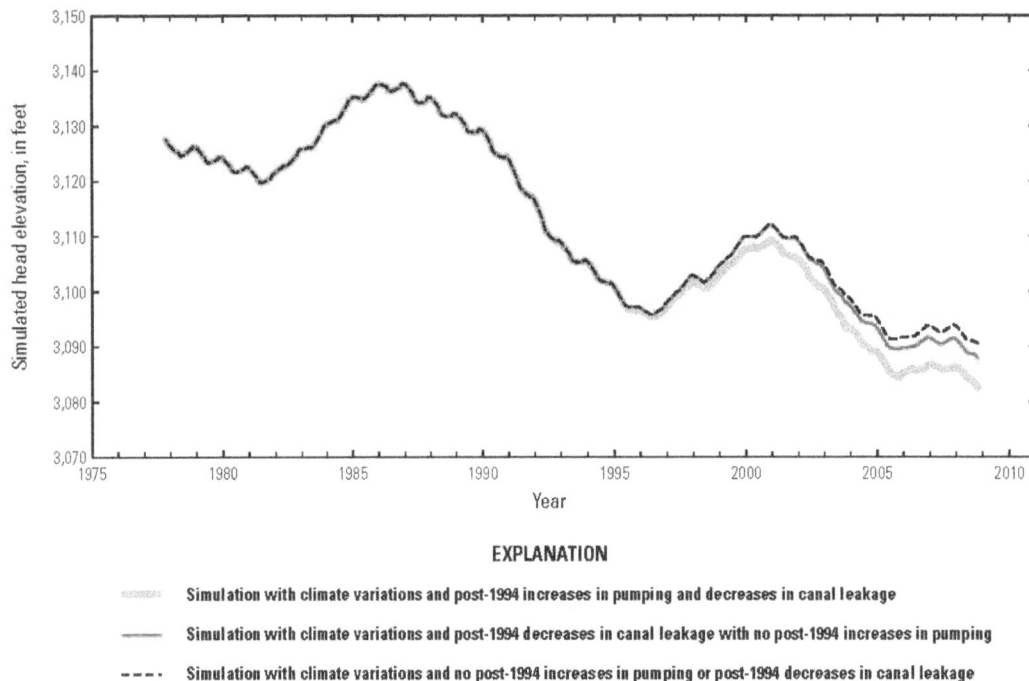

EXPLANATION

Simulation with climate variations and post-1994 increases in pumping and decreases in canal leakage

Simulation with climate variations and post-1994 decreases in canal leakage with no post-1994 increases in pumping

Simulation with climate variations and no post-1994 increases in pumping or post-1994 decreases in canal leakage

Figure 20.   Simulated head elevations in well 15S/10E-08ACD near Sisters, Oregon.

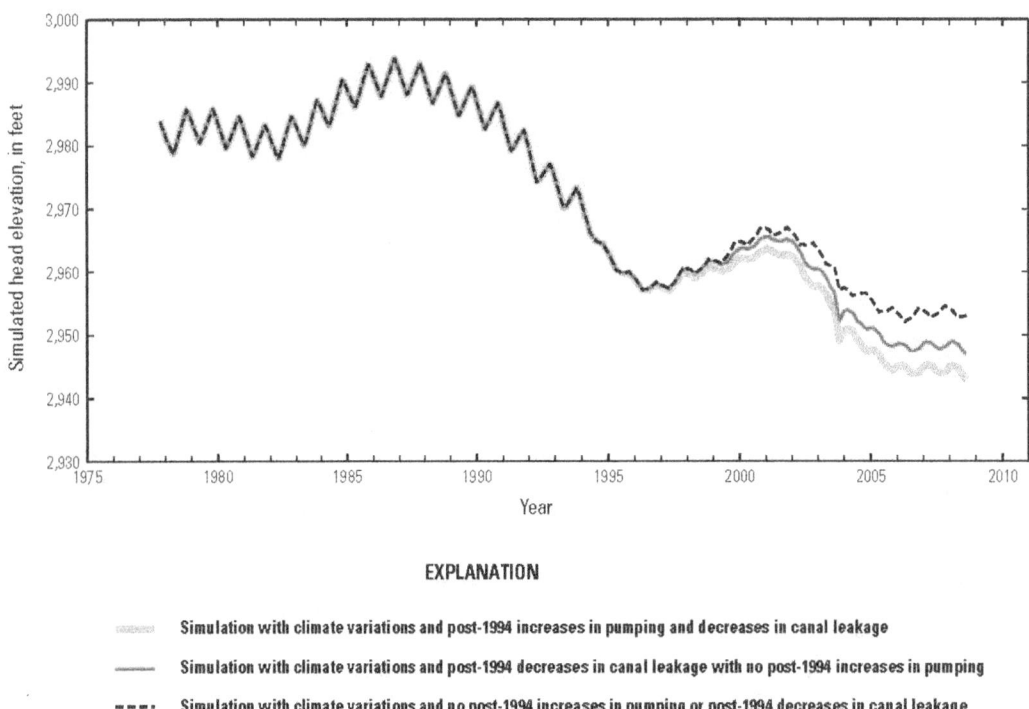

EXPLANATION

Simulation with climate variations and post-1994 increases in pumping and decreases in canal leakage

Simulation with climate variations and post-1994 decreases in canal leakage with no post-1994 increases in pumping

Simulation with climate variations and no post-1994 increases in pumping or post-1994 decreases in canal leakage

Figure 21.   Simulated head elevations in well15S/10E-02CDA east of Sisters, Oregon.

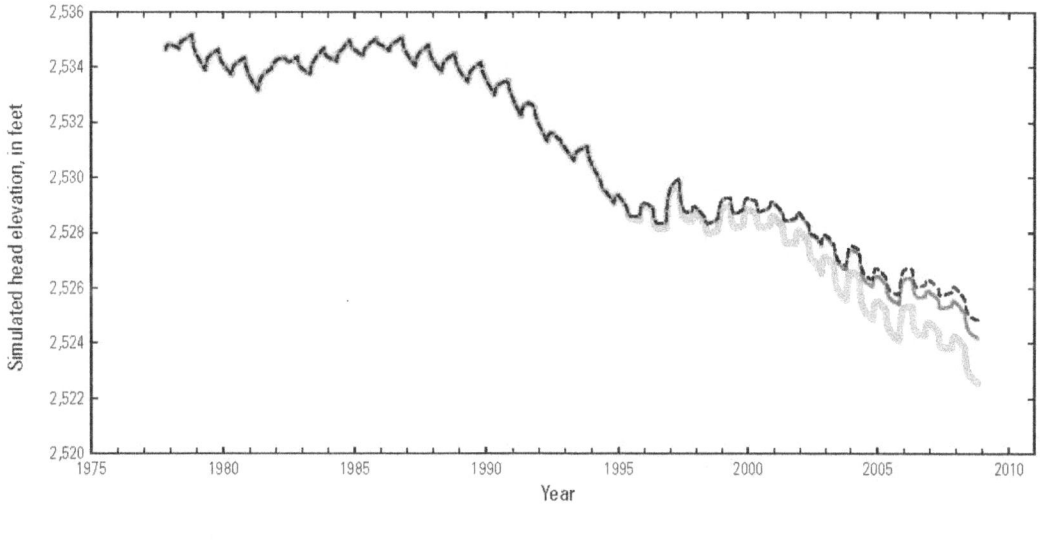

EXPLANATION

　　　　　Simulation with climate variations and post-1994 increases in pumping and decreases in canal leakage

　　　　　Simulation with climate variations and post-1994 decreases in canal leakage with no post-1994 increases in pumping

- - - -　Simulation with climate variations and no post-1994 increases in pumping or post-1994 decreases in canal leakage

Figure 22.　Simulated head elevations in well 14S/11E-01DDD1 near Lower Bridge, upper Deschutes Basin, central Oregon.

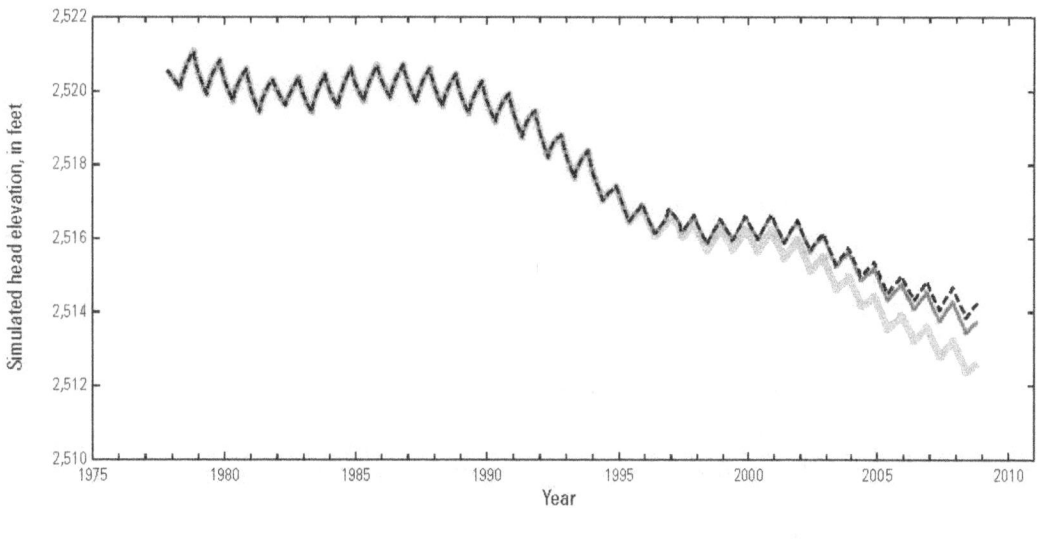

EXPLANATION

　　　　　Simulation with climate variations and post-1994 increases in pumping and decreases in canal leakage

　　　　　Simulation with climate variations and post-1994 decreases in canal leakage with no post-1994 increases in pumping

- - - -　Simulation with climate variations and no post-1994 increases in pumping or post-1994 decreases in canal leakage

Figure 23.　Simulated head elevations in well 14S/12E-02CCC near Lower Bridge, upper Deschutes Basin, central Oregon.

Cline Buttes to Redmond Area

Water levels in the area between Cline Buttes and Redmond have declined about 12–14 ft since the mid-1990s. As in the Lower Bridge area to the northwest, the decline has been more or less continuous since the mid-1990s. Declines persisted in the area west of Redmond, at a lesser rate in some wells, throughout the wet period in the late 1990s (figs. 24 and 25). Simulations show that about 60–70 percent of the measured decline in the area between Cline Butte and Redmond is likely due to climate influences, while 20–25 percent is due to increases in pumping, and 5–10 percent is the result of decreased recharge due to canal lining.

Water-level declines in the area between Cline Buttes and Redmond are about double those observed in the Lower Bridge area, even though the shapes of the trends are very similar. The declines in the Lower Bridge area may be attenuated because of proximity to the discharge area along the Deschutes River. Head-dependent flux boundaries, such as the gaining streams in the area, tend to buffer water-level fluctuations.

Redmond to Powell Butte Area

Water levels in the area between Redmond and the community of Powell Butte have had declines of about 13–14 ft since 1995. The decline has been persistent, but the rate of decline lessened during the wet period in the late 1990s (fig. 26). Like the Lower Bridge area, the Redmond to Powell Buttes area did not experience a water-level recovery during the late-1990s wet period, most likely because of the attenuation of the recharge pulse with distance from the Cascade Range recharge area. Simulations indicate that about 60–65 percent of the measured decline is due to climate, about 25–30 percent is due to post-1994 increases in pumping, and about 10 percent is due to decreases in recharge due to canal lining since the mid-1990s.

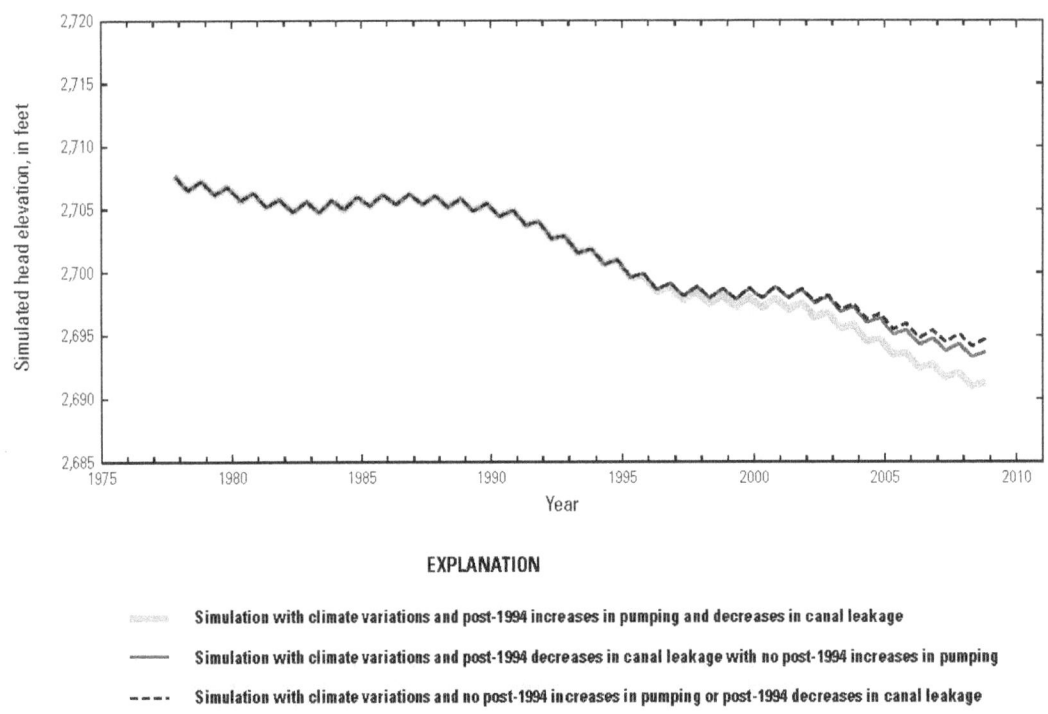

EXPLANATION

⎯⎯⎯ Simulation with climate variations and post-1994 increases in pumping and decreases in canal leakage

⎯⎯⎯ Simulation with climate variations and post-1994 decreases in canal leakage with no post-1994 increases in pumping

- - - - Simulation with climate variations and no post-1994 increases in pumping or post-1994 decreases in canal leakage

Figure 24.   Simulated head elevations in well 15S/13E-18ADD1 near Redmond, Oregon.

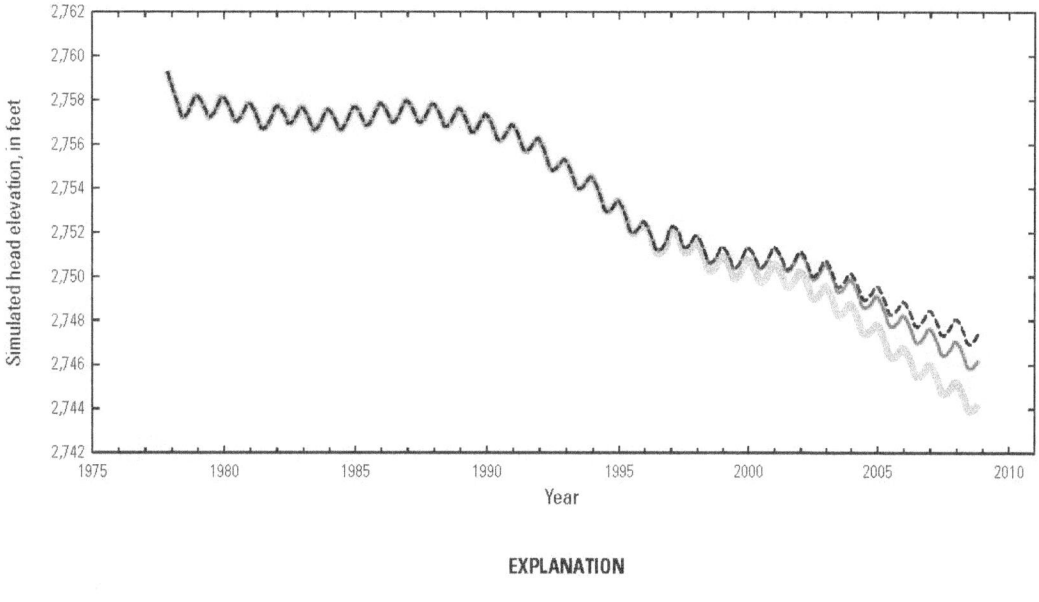

EXPLANATION

————— Simulation with climate variations and post-1994 increases in pumping and decreases in canal leakage

————— Simulation with climate variations and post-1994 decreases in canal leakage with no post-1994 increases in pumping

— — — Simulation with climate variations and no post-1994 increases in pumping or post-1994 decreases in canal leakage

Figure 25.   Simulated head elevations in well 15S/12E-14CDD west of Redmond, Oregon.

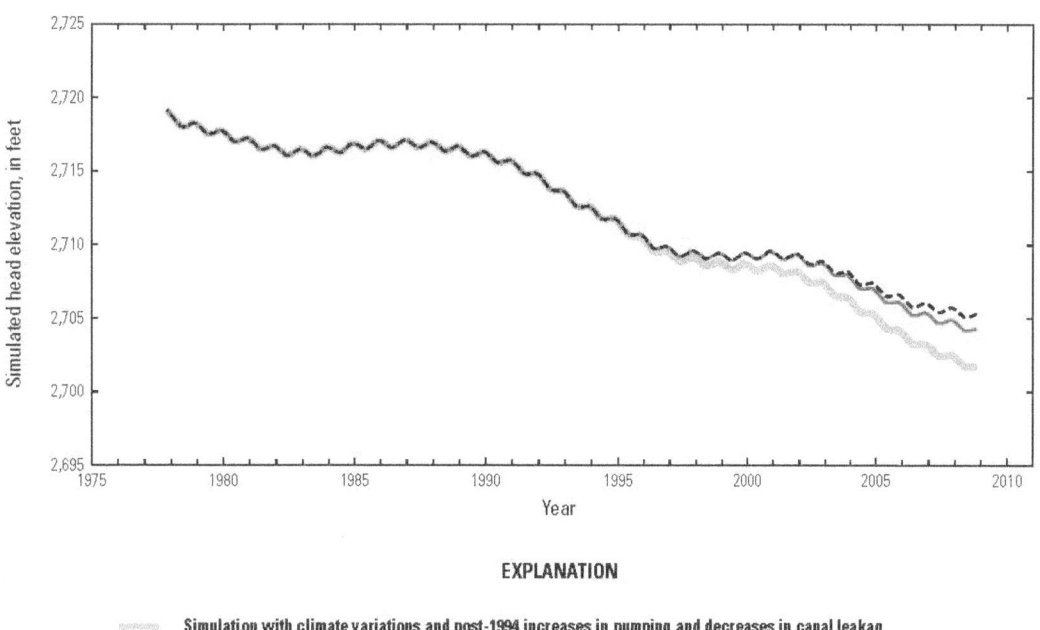

EXPLANATION

————— Simulation with climate variations and post-1994 increases in pumping and decreases in canal leakag

————— Simulation with climate variations and post-1994 decreases in canal leakage with no post-1994 increases in pumping

— — — Simulation with climate variations and no post-1994 increases in pumping or post-1994 decreases in canal leakage

Figure 26.   Simulated head elevations in well 15S/13E-21ADB1 near Redmond, Oregon.

Spatial Distribution of Pumping and Canal-Lining Influences

Maps showing simulated water-level changes resulting from post-1994 increases in pumping and decreased recharge due to canal lining can provide insights into the spatial distribution and magnitude of the impacts from these stresses. Such maps are created by subtracting 2008 water levels in each cell simulated in the base run from the 2008 water levels for the same cells from model runs during which post-1994 pumping or canal leakage were held at 1994 rates. Water-level changes resulting from decreases in canal leakage or due to increases in pumping vary with depth. Pumping impacts are largest at depths (and in strata) from which the water is being withdrawn. The effects of canal lining, in contrast, are most prominent at shallow depths closest to canals, and attenuate with depth.

The simulated water-level declines due to the post-1994 growth in groundwater pumping are generally centered on the area around Bend, Sisters, Redmond, and Powell Butte (fig. 27). In model layer 1, which corresponds to the upper 100 ft of the saturated zone (the zone beneath the water table), post-1994 pumping-related declines range from 1 to 5 ft over most of the populated part of the basin as of 2008, with local areas of up to 10 ft near centers of concentrated pumping near Sisters, Bend, and Powell Butte (fig. 27A). Deeper in the aquifer system, in model layer 3 (200- to 300-ft below the water table), the pumping-related declines are more evenly spread out, but still generally range from 1 to 5 ft, with some areas showing declines of 10–50 ft in the Bend area (fig. 27B). As described in preceding sections, simulated impacts from post-1994 pumping are consistent with measured water levels at observation wells. There are no monitoring wells in the Bend area in the areas where simulations show possible impacts.

Simulated water-level declines resulting from decreased groundwater recharge due to canal lining since 1994 range from 1 to 5 ft as of 2008, over a broad area encompassing Bend, lower Tumalo Creek, the Sisters area, and Redmond (fig. 28). Simulated water-level declines are as much as 68 ft in model layer 1, however, adjacent to canals with the largest reported decreases in leakage (fig. 28A). Simulated post-1994 water-level declines due to canal lining are more subdued, generally less than 10 ft, in model layer 3, with declines of up to about 15 ft near Tumalo Creek (fig. 28B). Simulations show that water-level declines up to 10 ft in adjacent to lined canals near Madras as of 2008.

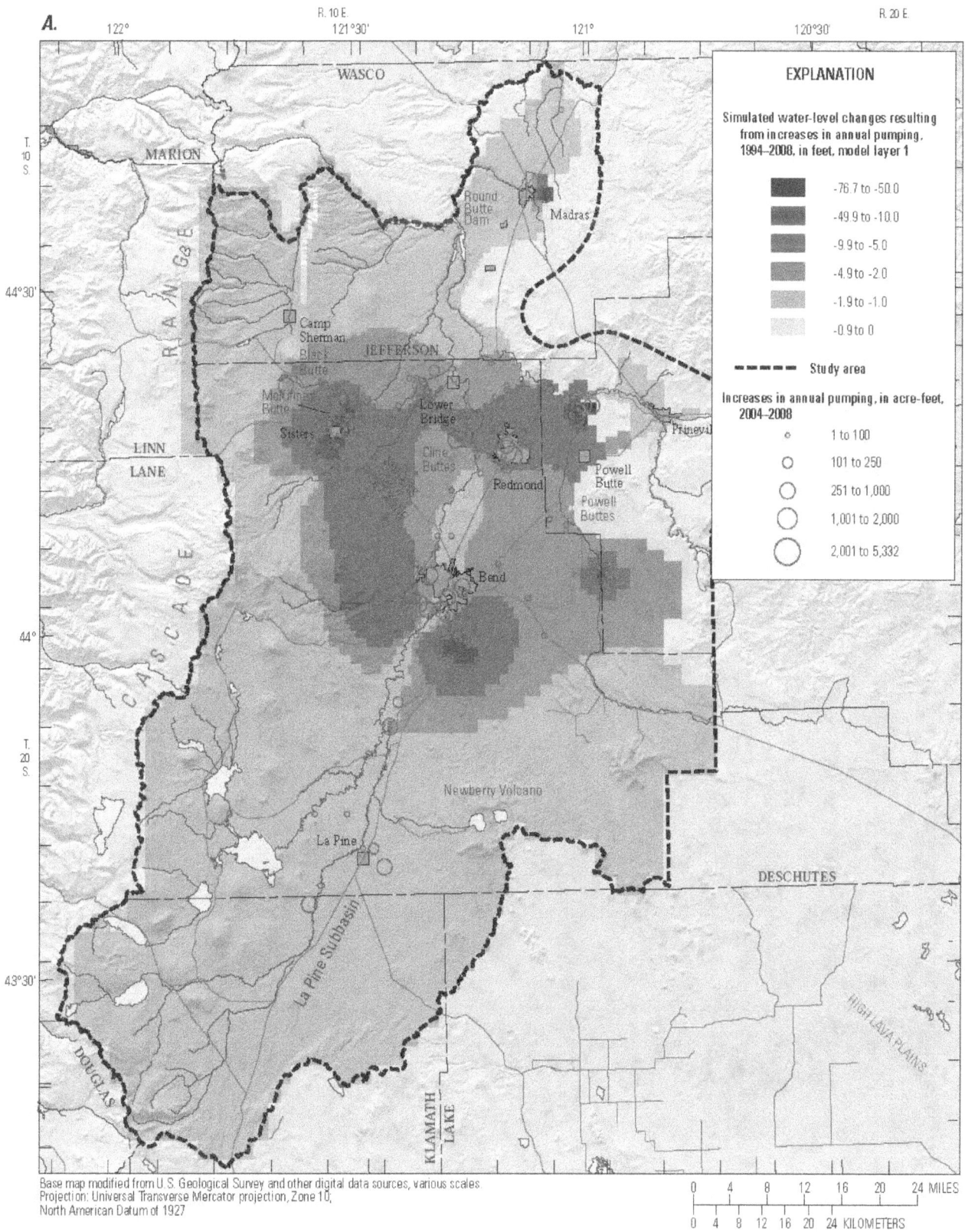

Figure 27. Increases in groundwater pumping since 1994 and simulated groundwater-level declines resulting only from post-1994 increases pumping in the upper Deschutes Basin, Oregon as of 2008. (*A*), model layer 1; (*B*), model layer 3.

Figure 27.—Continued

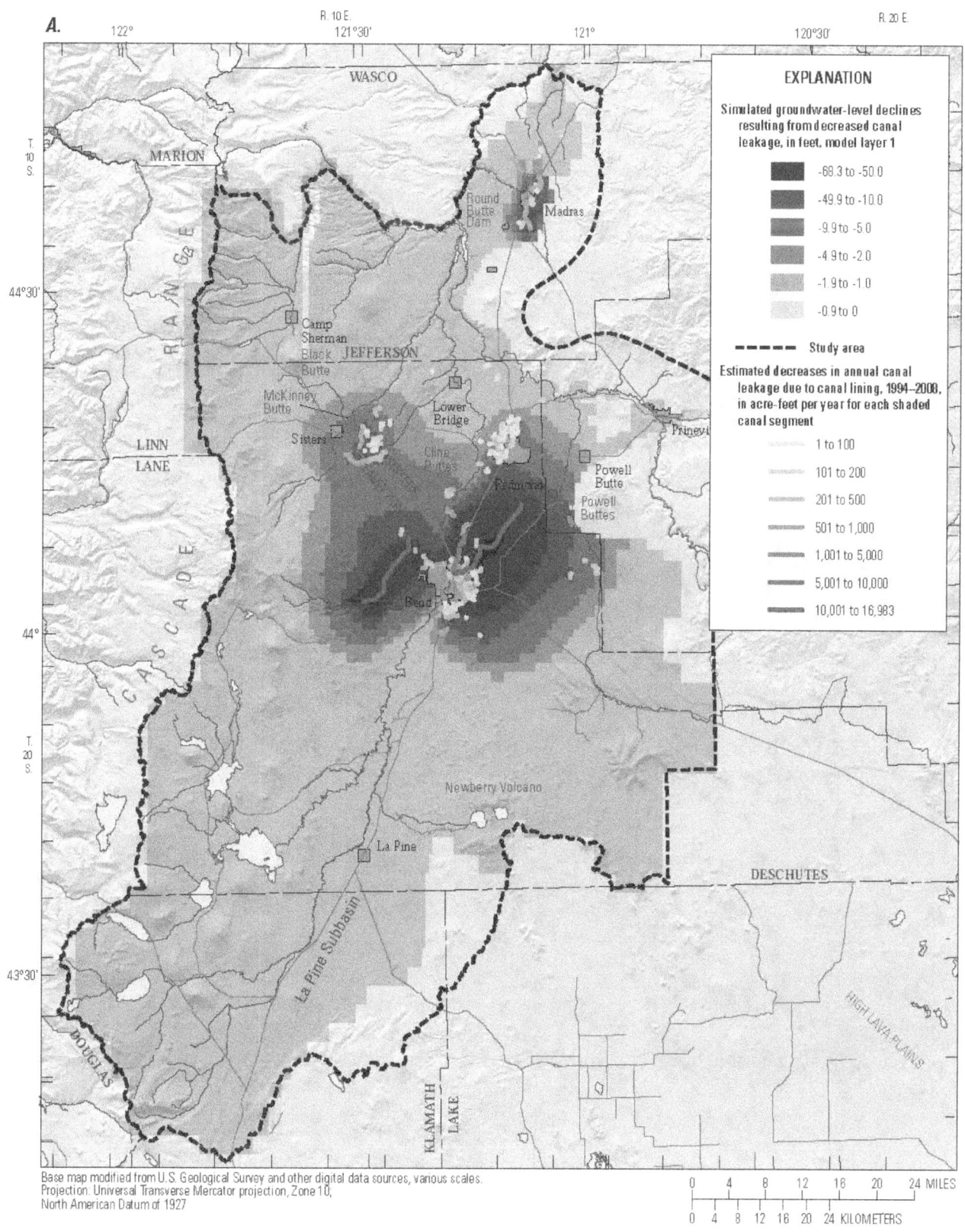

Figure 28.   Upper Deschutes Basin, Oregon, showing estimated decreases in annual canal leakage due to canal lining between 1994 and 2008, and simulated groundwater-level declines resulting from decreased canal leakage. (*A*) model layer 1; (*B*) model layer 3.

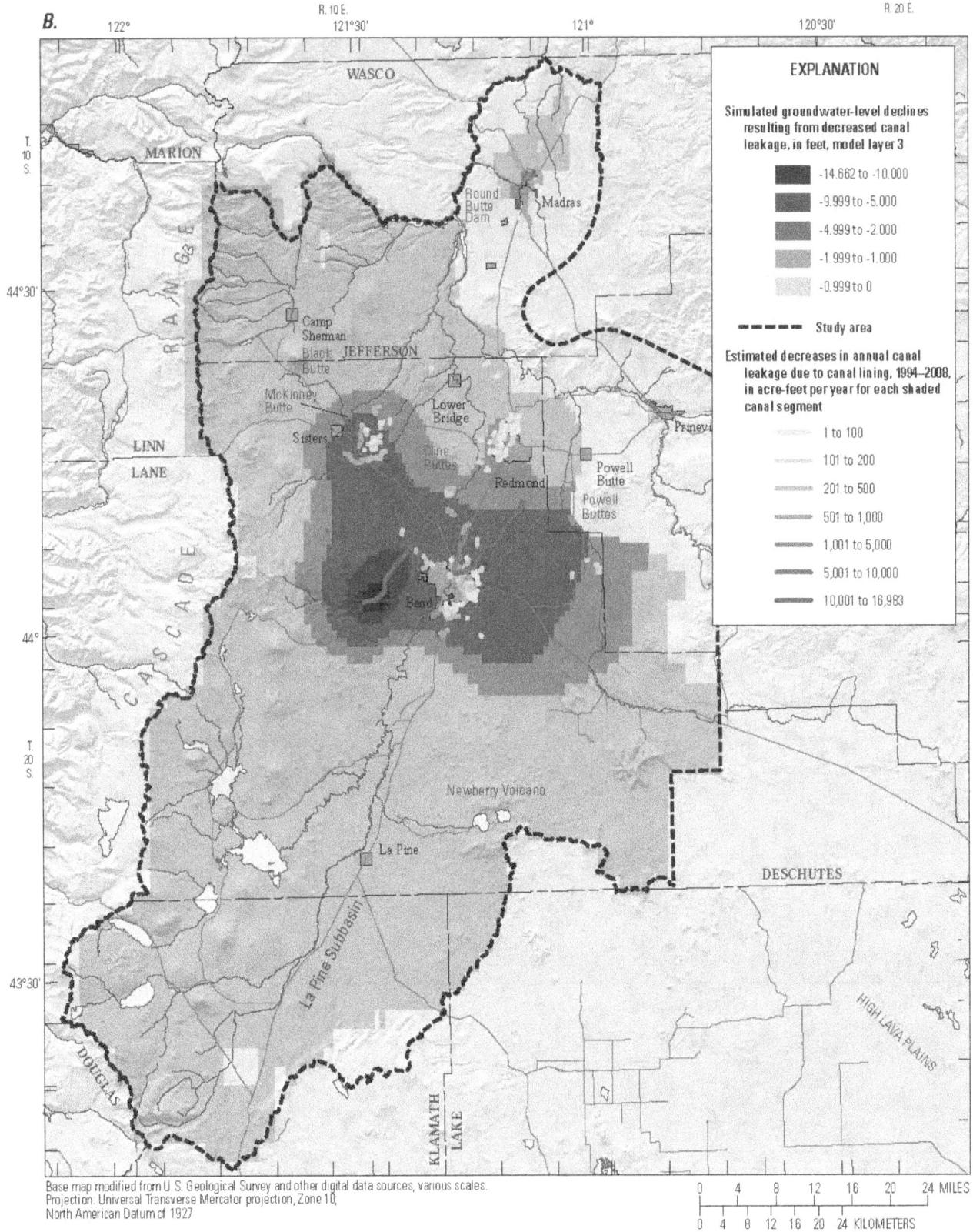

Figure 28.—Continued

Summary

Water levels in the central parts of the upper Deschutes Basin have declined by as much as 14 ft since the mid-1990s. The factors that affect groundwater levels in the basin and that are contributing to the measured declines include climate variations (both long-term trends and decadal cycles), increases in groundwater pumping, and decreases in artificial recharge due to lining of irrigation canals. The relative contribution of each of these factors has been evaluated using a groundwater-flow model. Water-level changes are dominated by climatic influences. In the central part of the basin, however, increases in groundwater pumping and decreases in recharge due to canal lining have significantly contributed to water-level declines.

The upper Deschutes Basin has experienced a general drying trend for the past several decades. The drying trend is manifested as a decrease in annual mean flow of many streams in the region, as well as decreases in groundwater discharge to spring-fed streams such as Fall River since 1950. Decreases in streamflow since the late 1950s in the Cascade Range and throughout the Pacific Northwest are well documented. Groundwater recharge calculated for this study using the Deep Percolation Model decreased about 25 percent between the 1979–88 period and the 1999–2008 period. A decrease in groundwater recharge is consistent with historical measurements that show that the discharge of spring-fed streams has also decreased in recent decades.

Superimposed on this long-term trend are seasonal variations and cyclic wet and dry periods (drought cycles) that occur on decadal time scales. The basin experienced a wet period from the mid-1990s to about 2000 that was followed by a dry cycle that lasted until about 2005. Climate conditions were normal or wetter than normal from 2006 to 2008. In response, measured groundwater levels in the Cascade Range, the principal groundwater recharge area, rose about 20 ft from the mid-1990s to 2000, and then declined a similar amount between 2000 and 2005. Water levels again rose as much as 10 ft between 2005 and 2008. The period of record for monitored wells in the Cascade Range is too short to discern the multi-decadal decrease observed in the streamflow and precipitation data.

The decadal climate cycles observed in the Cascade Range groundwater levels are increasingly attenuated toward the east in the central parts of the basin. Moving east from Sisters, for example, the decadal variations observed in the Cascades are more subtle, and the hydrographs are increasingly dominated by the multi-decadal declining trend. The short-term wet periods that result in marked rises in groundwater levels in the Cascade Range manifest in the interior parts of the basin as decreases in the rates of water-level declines.

Model analysis has provided insights into the relative contribution of climate variations, increased pumping, and increased canal lining on measured groundwater-level declines in the upper Deschutes Basin. Modeling has also provided insights into the geographic distribution of the response to these stresses. The effects of increased pumping and increased canal lining, as it turns out, are largely limited to the developed interior parts of the basin extending north-south roughly from Benham Falls north to Lower Bridge, and east-west from the area of Sisters to Powell Butte (the eastern extent of the model). Decreases in recharge due to on-farm losses (deep percolation of applied irrigation water) were included in the simulation analysis but were sufficiently small that they were considered negligible and not included in the discussion.

Water levels in the Sisters area rise and decline in response to climate in a manner similar to that observed in the Cascade Range. Despite longer term declines, 2008 water levels in the Sisters area were at or above the mid-1990s levels due to wet climatic conditions. West of McKinney Butte, water levels observed in the Sisters area have risen about 10 ft since 2005. Water-level trends in the Sisters area east of McKinney Butte, however, have remained flat, possibly reflecting the effects of the Sisters Fault Zone and larger distance from the principal recharge area in the Cascade Range. Simulations show that post-1994 canal lining and increases in groundwater pumping may be jointly responsible for head losses of approximately 6–8 ft in the Sisters area as of 2008.

The Lower Bridge area northeast of Sisters has experienced water-level declines of about 5 to 6 ft since the mid-1990s. Of these measured declines, about 3–4 ft, or 60–70 percent, can be attributed to climate, 1–2 ft, or about 20–30 percent, to post-1994 increases in pumping, and about 0.5 ft (roughly 10 percent) to canal lining. Water-level declines in the Lower Bridge area have been more or less continuous, with the rate of decline changing with wet and dry climate cycles.

The area extending from Cline Buttes through Redmond, and east to the community of Powell Butte, has seen groundwater-level declines of about 12–14 ft between the mid-1990s and 2008. Of this decline, 7–10 ft (or about 60–70 percent) can be attributed to climate, 2.5–3.5 ft (20–30 percent) to increases in pumping since 1994, and 0.5–1.5 ft (5–10 percent) to canal lining. As with the Lower Bridge area, water levels in Redmond and the surrounding area have exhibited a more or less continual decline, the rate varying with climate cycles.

Spatial analysis of simulation results shows that water level impacts resulting from post-1994 canal lining and increases in pumping extend from the Benham Falls area north to Lower Bridge, and from the Sisters area east to the community of Powell Butte. Outside of this general area of impact, the effects of pumping and canal lining generally are in the range of hundredths to tenths of a foot. Simulated water-level changes match measured declines reasonably well in the northern part of the area of impact, but there is a lack of monitoring in the southern part of the area with which to track water-level changes and verify simulation results.

Acknowledgments

The authors gratefully acknowledge contributions of individuals whose efforts contributed to this report. Critical public-supply water use data were provided by Patrick Griffiths, City of Bend; Jan Wick, Avion Water Company; and Pat Dorning, City of Redmond. Jon Haynes (USGS) compiled water use data and Scott Waibel (USGS) updated and ran the Deep Percolation Model. Dan Polette (USGS) and Ned Gates (OWRD) revisited several of the wells from the earlier study. Kyle Gorman and Jonathan La Marche of OWRD shared their considerable insights into the hydrology and changes in the basin, and provided valuable data on canal lining and land idling.

References Cited

Bauer, H.H., and Vaccaro, J.J., 1987, Documentation of a deep percolation model for estimating ground-water recharge: U.S. Geological Survey Open-File Report 86-536, 180 p.

Boyd, T.G., 1996, Groundwater recharge of the middle Deschutes Basin, Oregon: Portland, Oregon, Portland State University, M.S. thesis, 86 p.

Caldwell, R.R., and Truini, Margot, 1997, Ground-water and water-chemistry data for the upper Deschutes Basin, Oregon: U.S. Geological Survey Open-File Report 97–197, 77 p.

Gannett, M.W., and Lite, K.E., Jr., 2004, Simulation of regional ground-water flow in the upper Deschutes Basin, Oregon: U.S. Geological Survey Water-Resources Investigations Report 03-4195, 84 p.

Gannett, M.W., Lite, K.E., Jr., Morgan, D.S., and Collins, C.A., 2001, Ground-water hydrology of the upper Deschutes Basin, Oregon: U.S. Geological Survey Water-Resources Investigations Report 00-4162, 77 p. (Also available at http://or.water.usgs.gov/pubs/WRIR00-4162/.)

Hill, M.C., and Tiedeman, C.R., 2007, Effective groundwater model calibration: Hoboken, New Jersey, John Wiley and Sons, 455 p.

Luce C.H., and Holden, Z.A., 2009, Declining annual streamflow distributions in the Pacific Northwest United Stated, 1948–2006: Geophysical Research Letters, v. 36, L161401, doi:10.1029/2009GL039407, 2009, accessed January 5, 2013 at http://onlinelibrary.wiley.com/doi/10.1029/2009GL039407/abstract.

Manga, Michael, 1997, A model for discharge in spring-dominated streams and implications for the transmissivity and recharge of Quaternary volcanics in the Oregon Cascades: Water Resources Research, v. 33, no. 8, p. 1,813–1,822.

Mayer, T.D., and Naman, S.W., 2011, Streamflow response to climate as influence by geology and elevation: Journal of the American Water Resources Association, v. 47, no. 4, p. 724–738.

National Agricultural Statistics Service, 2007, Oregon and Washington Cropland Data Layer: National Agricultural Statistics Service, accessed January 5, 2013 at http://www.nass.usda.gov/research/Cropland/metadata/metadata_or07.htm.

Oregon Water Resources Department, 2013, Water level data and hydrographs: Oregon Water Resources Department, accessed January 8, 2013 at http://www.oregon.gov/owrd/pages/gw/well_data.aspx.

Sceva, J.E., 1968, Liquid waste disposal in the lava terrane of central Oregon: U.S. Department of the Interior, Federal Water Pollution Control Administration, Technical Projects Branch Report No. FR-4, 66 p., plus a 96 p. appendix.

Vaccaro, J.J., 2007, A deep percolation model for estimating ground-water recharge—Documentation of modules for the modular modeling system of the U.S. Geological Survey: U.S. Geological Survey Scientific Investigations Report 2006-5318, 30 p.

www.ingramcontent.com/pod-product-compliance
Lightning Source LLC
Chambersburg PA
CBHW081404170526
45166CB00010B/3207